# How to Read the Weather

## STORM DUNLOP

🍂 **National Trust**

First published in the United Kingdom in 2018 by
National Trust Books
43 Great Ormond Street
London
WC1N 3HZ

An imprint of Pavilion Books Company Ltd

ISBN 978-1-91135-824-4

A CIP catalogue record for this book is available from the British Library.

10 9 8 7 6 5 4 3 2 1

Reproduction by Mission Productions Ltd, Hong Kong
Printed and bound by 1010 Printing International Ltd, China
This book can be ordered direct from the publisher at
www.pavilionbooks.com. Also available at National Trust shops,
including www.nationaltrustbooks.co.uk

**Pages 4–5** Snow at Hardwick Hall, Derbyshire.

# Contents

# 1. Weather Fundamentals

# The Local Weather Forecast

Having the 'right' weather is important to so many of our activities, yet sometimes it seems the weather forecasters are always getting it wrong. Occasionally their forecasts are wildly inaccurate, but more often there are little differences that can mess up our plans: rain when none was expected spoils our day at the beach, or a fine, sunny day when the local forecast had said showers means we'll have to water the garden after all.

To be fair, weather is one of the most complex phenomena that we experience, and creating correct forecasts is an exceptionally difficult task. People have been trying to do it since at least the Babylonians, who in around 3000BC were trying to predict weather using astrology – the position of the moon and the planets in the sky – and weather lore. Even today,

**Above** A glorious day at Felbrigg Hall, Norfolk.

weather forecasting is possibly the most complex scientific problem that exists. But it has improved significantly in recent decades, with modern supercomputers used by the majority of major meteorological offices. In fact, the overall pattern of their weather predictions is essentially correct. It's in the details of local weather that problems still arise. Despite all the major advances in forecasting, as yet none of the models are able to account for the 'minor' local variations that can put paid to our plans for the day.

Predicting local weather is a minefield. Knowledge of the features that may affect local weather – hills or mountains, bodies of water, and also human activities – may help to supplement the official forecasts, but these weather 'complications' may themselves introduce considerable variations. And it is impossible to focus on local weather without also bearing in mind the global influences at work. Even the less changeable local areas are subject to the same global processes that govern the weather as a whole.

Britain's weather is subject to a wide variety of influences. It has a maritime climate, from being located on the western edge of Europe, bordered by the Atlantic. A warm ocean current, the North Atlantic Drift, brings moist, warm air. As a result it has generally mild weather in both summer and winter, but it also has highly variable local weather.

# What is Weather?

Weather is the condition of the atmosphere, which we describe colloquially as sunny, rainy, snowing, hot or cold. The basic factors that make weather are temperature and pressure, as well as the behaviour of water in the atmosphere. It pays to understand all these more fully, and how they interact.

The Earth's weather systems are ultimately driven by the difference in heating between equatorial regions and the poles. Radiation from the Sun heats the Earth unevenly, and any given area at the poles receives far less heating than a similar area at the equator, which creates strong temperature differences. Warm air at the equator expands and rises, creating low pressure at the surface while the colder air at the poles contracts, creating higher pressure at the poles. The imbalance between the temperature and pressure at different latitudes creates the overall circulation.

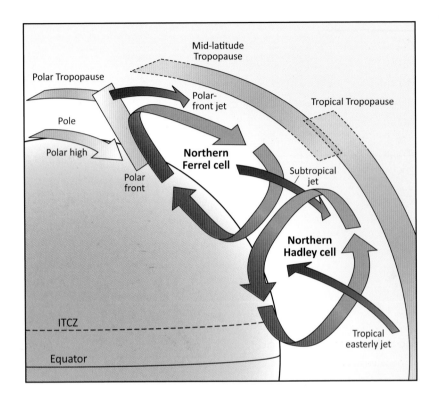

# Winds and Circulation Systems

In the early days of studying the weather, people thought that warm air rising in the tropics flowed north and south, then descended in the polar regions as cold air, after which it flowed back to the equatorial region – all within a single circulation system. But it wasn't long before meteorologists realised that there are actually three major circulation cells that form wind belts around the Earth: from the equator to the mid-latitude high-pressure areas (known as the Hadley cell), a mid-latitude cell (the Ferrel cell), and a polar cell. These three cells exist in both northern and southern hemispheres with an accompanying pattern of high- and low-pressure areas. There will be more about them in Chapter 3. The way these systems move and govern the weather we experience is determined by the overall pattern of the winds.

## Expansion and Contraction

Both expansion and contraction of air, and how they relate to temperature, are important in understanding the weather and, for example, how clouds form. Air cools as it expands and warms as it is compressed. A couple of everyday examples make this clearer. When you pump up a tyre, the pump becomes hot as the air is compressed through the pump into the tyre and its kinetic energy (its heat) increases. But when you let air out of a tyre, it feels cold as pressure is released and the air expands – in the same way, the air or propellant from an aerosol can feel cold when it is released from the pressurised container.

# In the Doldrums

Air rising in the equatorial region creates a belt of low pressure at the surface, known to sailors as the Doldrums, where winds are light and greatly variable or even non-existent. This air flows north and south high in the atmosphere, finally descending at latitudes of approximately 30°N and 30°S. As air descends it becomes compressed and warms (more detail on this all-important feature later), producing high-pressure regions at the surface. In this particular case the high-pressure belts created are known as the 'subtropical anticyclones' or 'subtropical highs'. The descending air becomes dry and hot, and the Earth's major deserts are located at these latitudes. In the northern hemisphere there are the Sahara and Arabian Deserts, as well as the desert areas in North America. (The Sahara is the source of the hottest air that reaches Britain and the western part of Europe.) In the southern hemisphere there are smaller land areas compared with the vast extent of the Southern Ocean, so this means there are smaller desert regions, most notably the Namib and Kalahari Deserts in southern Africa, and the Great Australian Desert.

# Trade Winds

Surface air flows out from the subtropical high-pressure belts and much of it returns back towards the equatorial regions, setting up a consistent flow. This forms what are known as the trade winds, which blow mainly from the north-east in the northern hemisphere and from the south-east in the southern. The northern and southern trade winds meet up in the equatorial region, and this important boundary is known as the Intertropical Convergence Zone (ITCZ). A bit of a mouthful to remember, but actually you can quite easily see it on satellite images of the Earth as a distinct line of clouds. The trade-wind zones typically show shallow cumulus clouds, which are limited in their upward growth by an inversion (see page 25) created by air subsiding above them.

## Where the Wind Blows

You might assume that the 'trade' in trade winds refers to their importance to commerce. In fact it originally came from an older usage of the word, 'track' or 'habitual path', to mean the winds were consistent. Our later use of 'trade' as commerce comes from the same root, as the 'habitual path of business'. Winds, such as the north-east trades or the westerlies, are always described by the direction from which they originate or 'blow from' – more on that later.

# Westerlies and Easterlies

While trade winds circulate in the subtropical to equatorial regions, some of the air from the subtropical highs flows towards the poles rather than the equator to become the winds known as westerlies. These dominate the weather over most of the middle latitudes of the Earth, in the temperate zones where most of the population lives.

## The Roaring Forties

In the southern hemisphere, the region of the persistent, very strong westerlies came to be known to sailors as the Roaring Forties, because they were at a latitude around 40–50°. In the days of sail, they were of great assistance to ships sailing from Europe to the East Indies and Australia.

There are a number of semi-permanent high- and low-pressure centres around the Earth, both of which are known as 'centres of action'. In general forecasting, high-pressure centres are known as anticyclones, whereas low-pressure centres – technically 'cyclones', are known as depressions when they are the centres of active weather systems.

## Seasonal Winds

At the poles, the air is naturally cold and creates shallow high-pressure regions called the 'polar highs'. The cold air spreads out from these at the surface, heading generally towards the equator. This flow of air produces the polar easterlies. Where this cold air meets the warm air flowing out of the mid-latitude high-pressure regions, these very important boundaries are known as the polar fronts. The latitude of these polar fronts in the northern and southern hemispheres is very variable – it accounts for the extreme variations in the weather that are experienced in the middle latitudes of the Earth – but they are generally located at approximately 40–50°N and S.

# Which direction?

**In the northern hemisphere, air flows out of high-pressure anticyclones in a clockwise direction and out of low-pressure cyclones (depressions) in an anticlockwise direction. The directions are reversed in the southern hemisphere.**

The general pattern of pressure centres such as these tends to shift north and south slightly with the seasons. The air flowing out of an anticyclone gives rise to surface winds. This is most noticeable in the northern hemisphere, where the high-pressure anticyclone of the Azores High (also known as the Bermuda High) tends to become particularly marked and stronger in summer, when it extends its high pressure towards the British Isles and Europe. A similar high-pressure zone intensifies over the Pacific Ocean in summer (the North Pacific High).

There are also certain semi-permanent low-pressure regions, the most significant in the northern hemisphere being the Icelandic Low and the Aleutian Low, that are particularly marked in winter. Their nature is different from the high-pressure anticyclones. The persistent surface low pressure is from the low-pressure systems (known as extra-tropical cyclones or depressions) that often pass across them, rather from their being a semi-permanent feature created by the overall circulation.

Some features change more significantly between winter and summer. In the northern hemisphere, the winter sees a giant, cold, high-pressure system over northern Asia (the Siberian High) and cold, dry air flows out from that region. But in summer that area decays, and a low-pressure region (the Asian Low) develops instead, slightly further south. How does that happen? It's partially caused by intense heating over the Thar Desert in western India, which creates low surface pressure. When that happens, air flows inwards towards this low-pressure area in southern Asia. This alternation between high pressure in winter and low in summer also reverses the pattern of winds in winter and summer – bringing dry, cold northerly winds in winter and warm, moist south-westerly winds in summer. A similar reversal also occurs over central Africa. Wind patterns that reverse in this way are known as monsoons. They carry the torrential rain that the summer south-west monsoon wind brings in Asia, to India and nearby countries, providing a much-needed end to the intense heat of preceding months.

**Left** A wind-swept wild pony at Carneddau and Glyderau in Gwynedd, North Wales.

# The Coriolis Effect

When it was first understood that the circulation of air at the surface of the Earth was driven by the temperature difference between equatorial regions and the poles, people assumed the course of the winds would be generally in a north–south direction. But the discovery of the relatively constant trade winds – north-easterlies in the northern hemisphere and south-easterlies in the southern – showed this wasn't the case. There were also the strong westerlies in the middle latitudes.

The reason that winds don't follow a north–south direction is due to the rotation of the Earth. And this is also why, even locally, and well away from the equator, the wind at some distance from the surface flows in a different direction to the winds nearer the ground. The

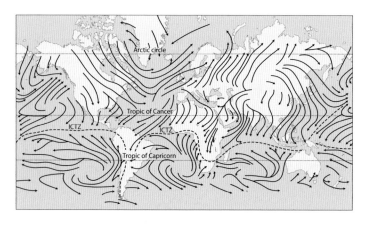

Global winds in January.

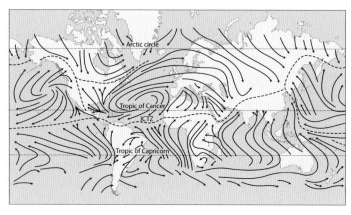

Global winds in July.

# Parcels of Air

Meteorologists commonly refer to a 'parcel of
that has all the properties of the atmosphere bu
discussing how air behaves in response to certain

rotation of the Earth deflects winds in
the northern hemisphere to the right of
their original direction and those in the
southern to the left. This is known as
the Coriolis Effect.

To explain in more detail, the
strength of the Coriolis Effect depends
on two factors: the latitude and the
velocity of the moving air. At the
equator, the Earth is rotating towards
the east at a speed of 40,074km (24,901
miles) in 24 hours – about 1,670km
(1,038 miles) an hour. By contrast, at
the poles there is no eastwards motion,
merely one rotation of the Earth over
24 hours. Imagine a parcel of air at the
equator that starts to move north or
south. As it moves towards the pole,
this air retains its original eastwards
velocity, but the surface beneath it is
progressively moving eastwards at a
lower velocity. The air will no longer
flow directly towards the low-pressure
centre, under the influence of the
pressure gradient. Instead it will be
deflected, towards the right in the
northern hemisphere and towards the
left in the southern. Going the other

way, imagin
(which has nc
starts to move
It will move ovei
moving areas of ti
behind' the surface
also appears to be de
(in the north) or to the ,

The most important e
the air (not to mention on
moving objects, such as artil.
the force that acts in a horizont.
– that is, parallel to the Earth's s
often known as the Coriolis Force

At the equator there is no Corik
Effect, but it increases towards the pc
In addition, the force increases with th
velocity of the wind: the stronger the
wind, the greater the deflection. This
has an important effect. Friction at the
surface reduces the velocity of the wind,
meaning that at the surface the Coriolis
Force (and the change of direction)
is also reduced. This Coriolis Effect
results in the formation of the trade-wind
easterlies on either side of the equator,
the strong westerlies in the middle
latitudes, and the polar easterlies.

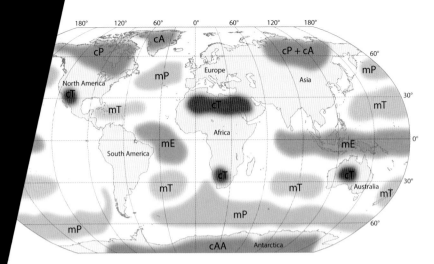

# r Masses and Fronts

rge masses of air flowing out from igh-pressure regions may stagnate n one location for a long time (weeks or even months) and take on certain characteristics of the underlying surface, in particular temperature and humidity. These areas are known as 'source regions'. A source region may be hundreds or even thousands of kilometres across, but its depth tends to depend on where the air mass forms. Cold air masses tend to be shallow, perhaps 1,000m (3,280ft) deep, but warm air masses are usually much deeper. They may grow to the whole depth of the troposphere (the lowest region of the atmosphere, from the Earth's surface to about 6–10km (3¾–6¼ miles) above it at the lower boundary of the stratosphere). In warm air masses, atmospheric circulation (convection) causes extensive mixing, which tends to equalise the temperature and humidity throughout the deep column of air.

Air masses are classified broadly as being of continental (c) or maritime (m) origin, depending on whether they are over land or sea. Continental air masses are dry while maritime masses are humid. Colder air masses are termed Arctic or Antarctic (A) or Polar (P) and warmer air masses are termed Tropical (T) or Equatorial (E). Equatorial air is always maritime in nature (never continental), so is always hot and humid.

**Above** The major source regions from which air moves around the world.

# THE MAIN TYPES OF AIR MASSES

| | | |
|---|---|---|
| Continental arctic | cA | extremely cold and dry |
| Continental polar | cP | cold and dry |
| Continental tropical | cT | hot and dry |
| Maritime arctic | mA | extremely cold and humid |
| Maritime polar | mP | cold and humid |
| Maritime tropical | mT | warm and humid |
| Maritime equatorial | mE | hot and humid |

The abbreviations are used by the UK Met Office and are commonly found in reference works. Sometimes the order of the words and letters is reversed, so that maritime tropical (mT), for example, is shown as tropical maritime (Tm).

Western Europe and Britain are primarily affected by four air masses – two maritime, and two continental. The maritime polar (mP) air mass has cold, humid air from the North Atlantic area, and the maritime tropical (mT) air mass has warm humid air originating in the subtropical anticyclone over the southern North Atlantic, close to the Tropic of Cancer. The continental polar (cP) air mass brings cold, dry continental air that comes from two source regions, over northern Canada and northern Eurasia, and the continental tropical (cT) air mass arises from the subtropical anticyclone over northern Africa.

The location and size of the source regions varies between winter and summer. As an example, in summer in the northern hemisphere, the Azores/Bermuda High – the source of the maritime tropical (mT) air – expands in area and sends ridges of high pressure (and warm moist air) over Western Europe. A similar source region (the Pacific High) develops over the North Pacific. The corresponding source regions for cold polar air shrink, in the case of the winter Siberian High disappearing completely.

There are other air masses that may influence Britain and Western Europe on occasion, with incursions of much colder, arctic air (A). These may be of maritime or continental origin. The source region for the dry, extremely cold continental arctic (cA) air is the ice-covered region of Arctic Canada or, less often, northern Siberia. Maritime arctic (mA) air usually originates in ice-free areas of the high Arctic Ocean, north of Scandinavia.

As an air mass moves across the globe, its characteristics change from those that it acquired over its source region. Dry continental air passing over water will become more humid, for example. There may be variations in temperature and humidity of the air that flows from a single source region, depending on the direction it takes. Hot, dry air from the Sahara, for example, may become cooler and more humid as it flows west over the cold waters of the Atlantic Ocean, whereas air from the same region flowing east over Egypt and the Arabian Desert remains hot and very dry. Similarly, maritime tropical air may pick up less moisture if it flows over cold seas, when compared with air flowing out of the same source region, but passing over areas with higher sea surface temperatures.

# Charting the Winds

## Atmospheric Pressure and Weather Forecasting

Air masses that are different in temperature and humidity are accompanied by different kinds of weather at the surface, but what effect do they have on it? One noticeable effect is their influence on local wind directions and speeds.

While the name may be unfamiliar, 'synoptic charts' of surface pressure will be familiar to most people, and are generally the most useful charts shown in TV weather forecasts. They show the pattern of surface pressure (and thus winds) and the isobars – lines joining points of equal pressure – at a specific time. These are used as the basis for forecasts of forthcoming weather, and are called 'synoptic' because they summarise actual observations obtained at the same time over a wide area. The UK Met Office issues these as 'analysis charts', showing the observed state of the weather, along with forecast charts up to five days ahead. Forecast charts of this kind are, naturally, subject to some uncertainty as actual conditions change and are normally revised at intervals of 12 hours following the analysis chart.

Anyone old enough to remember having a barometer in the hall, or perhaps at their grandparents', will know that atmospheric pressure was once measured by the height of mercury in the column, measured in inches. Today, most of us are more familiar with pressure expressed in millibars (mb). This is how it is usually expressed in weather forecasts. Meteorologists tend to use another unit, the hectopascal (hPa), which has more scientific precision. The average pressure at the surface of the Earth is defined as 1013.25hPa. Since one millibar (1mb) is precisely equal to one hectopascal (1hPa), the same figure may be used to express pressure in either unit.

# Beyond the Surface

At the Earth's surface, friction reduces the speed of the wind and the Coriolis deflection is less, so air flows across the isobars. Because of surface friction the angle at which the air flows across the isobars (rather than along them) varies, depending on the nature of the surface. Over the sea the angle is 10–20°, but over the land this increases to 25–35° because the wind is slowed.

Up in the 'free atmosphere', beyond the Earth's surface where friction plays a part and the Coriolis Effect has more effect, the air actually flows along the isobars, at right angles to the pressure gradient (the change in pressure over distance), and thus around areas of both low and high pressure. Technically, this freely flowing wind is known as the geostrophic wind. The height at which surface friction is no longer felt depends on the nature of the surface, but is generally taken to be between 500 and 1,000m (very approximately 1,500 to 3,000ft).

Forecasters are able to use charts showing the pressure distribution at height, where the pattern of isobars may be considerably different from the pattern at the surface. Although

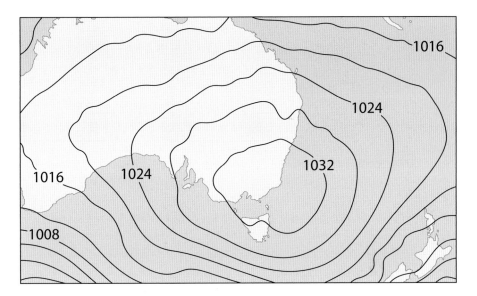

such charts are not readily available to the general public, high clouds (cirrus clouds) often provide an indication of the direction of upper winds. When this direction differs greatly from that of the surface winds – the wind directions are 'crossed' – this strongly indicates forthcoming changes in the weather. There is more about this in Chapters 2 and 3, in relation to jet streams and how they influence depression systems.

This difference in direction between upper and surface winds explains why clouds often seem to be moving on a slightly different path to the wind that you feel on the ground. Clouds do not need to be very high to be in the free atmosphere, where the (geostrophic) wind is blowing along the isobars. If you turn around and face the wind, there may be blue sky directly ahead but rain

clouds could be creeping up on you from slightly to the right (in the northern hemisphere) of the surface wind.

The effect of friction is not limited to areas of low and high pressure. A change

## Isobars and Pressure Gradients

Isobars are lines of constant pressure (think of them a bit like contour lines on a map). The pressure gradient is the change in pressure measured over a given distance, and acts at right angles to the isobars, from high pressure to low. If isobars are packed close together the gradient will be higher, and vice versa.

in an anticlockwise direction. When the wind direction changes in the opposite direction, clockwise, say from south-east to south, it is known as veering.

The location of nearby centres of high or low pressure obviously affects the local winds. But, if you think about it further, the air flowing out of those centres of high pressure – anticyclones – must itself have come from somewhere. In fact, it has descended from upper levels. Conversely, air flowing into the centre of a depression cannot accumulate indefinitely and must leave somehow. Actually, it ascends. Both these effects have important consequences for the weather and indeed are largely responsible for the overall weather patterns that we experience.

in the amount of friction caused by the surface will cause a change of wind speed and direction. The most obvious example is when the wind passes from the sea onto the land. The additional friction caused by the rough surface of dry land causes the wind to slow. That in turn decreases the Coriolis Force, so the wind adopts a more direct path from high pressure to low – it is deflected less to the right (in the northern hemisphere). To use the nautical term, the wind 'backs', or changes direction

**Above** Broken cumulus clouds behind heavier shower clouds.

# Low on the Left, High on the Right

There is a simple trick to finding out where your local centres of low and high pressure lie, using what is known as Buys-Ballot's Law – the Dutch scientist who first popularised it. If you are in the northern hemisphere, stand with your back to the wind and the low-pressure centre will be on your left. It's easy to remember as 'Low on the Left'. (Because surface friction over land changes the angle at which air flows across isobars, the actual centre of a low will be further forward.) And, correspondingly, high pressure is on your right. Reverse the directions if you are in the southern hemisphere.

# The Atmosphere

While most of the weather that affects us occurs in the lowest layer of the atmosphere, it's important to have some idea of what happens in the higher layers, especially as it's sometimes possible to see some interesting, but rare, clouds within them. It's also helpful to be able to understand some of the terms that you might hear used in weather forecasts. The influence of the jet streams (which are often mentioned) are also significant in the development of weather systems, and the clouds within them may sometimes provide a significant indication of how the weather may change.

As we've seen, the pattern of winds that blow horizontally across the globe is largely because of pressure changes and the pressure gradients that they produce. Such winds are known as 'gradient winds'. But vertical motion is also very important in governing the weather, on both global and local scales. The strength and development of depression systems (which usually bring great changes in the weather during a day), and also the growth of clouds and the weather that accompanies them, are largely caused by vertical motion.

# The Layers of the Atmosphere

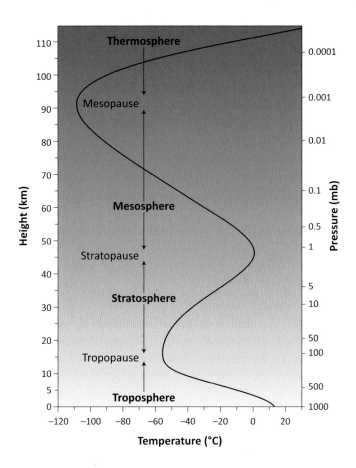

Almost all of the weather that affects the surface of the Earth occurs in the lowermost layer of the atmosphere, the troposphere. There is some influence from the next higher layer, the stratosphere, where clouds are rare. And even higher up comes the mesosphere and, on the edge of space, the exosphere. These layers are not that significant for weather, apart from the fact that some rare, interesting clouds are sometimes seen in the stratosphere and the mesosphere.

# The Tropopause and Stratopause

The different layers in the atmosphere are defined by the way in which temperature changes with height. Like everything else on the Earth, air is governed by the planet's gravity. As a result, and in general terms, air pressure is greatest at the surface and declines upwards towards the vacuum of space.

When pressure declines, air expands and cools. So it might be thought that temperatures would decrease steadily from the ground to space. But this is not the case. There is indeed an overall fall in the lowermost layer, the troposphere, but there may be levels within it (known as inversions) at which the temperature actually increases. These have an important effect on the development of clouds. There is always a major inversion at the top of the troposphere, known as the tropopause, where the temperature starts to rise (or remains constant over some kilometres). The tropopause forms a boundary between the troposphere and stratosphere and is a very important division. Its presence limits the upward growth of even the most vigorous clouds, the cumulonimbus. These clouds (described in more detail later) often develop flattened tops as they reach the tropopause, and thus give a visual indication of its location. You have probably noticed these kinds of flattened 'anvil-shaped' cloud forms, especially around the time of summer storms.

The height of the tropopause depends strongly on latitude. It may be as high as 16–18km (10–11⅕ miles) over the equator and no more than 7.5–9.5km (4²/₃– 6 miles) (or even lower) at the poles. It also tends to be higher in summer than in winter. The typical temperature at the tropopause is -55°C (-67°F). Above that height, in the stratosphere, the temperature rises through the absorption of energy from the Sun by ozone in the ozone layer, which lies between about 10 and 50km (6¼ and 31 miles) in altitude. It reaches about 0°C (32°F) at an altitude of 50km (31 miles). This level, the top of the stratosphere, is known as the stratopause.

Above the stratopause the temperature begins to fall throughout the mesosphere, reaching the lowest point in the atmosphere (about -163 to -100°C/-261°F to -148°F) at the mesopause. This lies at different heights in summer (about 100km/62 miles) and winter (approximately 86km/53½ miles). Beyond the mesopause lies the exosphere, where the atmosphere fades away into space.

# Ozone and the Ozone Layer

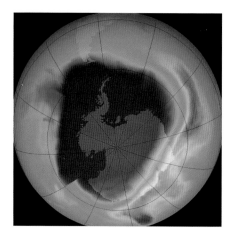

Ozone exists at three different levels in the atmosphere. At ground level, ozone is primarily produced by interactions between volatile organic compounds (VOCs), such as vehicle exhausts and industrial emissions, and nitrogen oxides, in the presence of heat and sunlight (which causes a photochemical reaction). At this level it is a very harmful gas. A certain concentration of ozone is found in the upper troposphere (at altitudes of 9–13km/5½ miles), where it is mainly produced by lightning and aircraft exhausts. The most significant concentration occurs in the stratospheric ozone layer, which is actually protective of the Earth rather than harmful. Here, ozone is created by the action of ultraviolet light from the Sun, which splits oxygen molecules ($O_2$), into individual oxygen atoms, which then combine to form molecules of ozone ($O_3$).

By absorbing the ultraviolet radiation, the ozone in the upper layer shields the Earth's surface from injurious UV radiation that, among its other effects, increases the incidence of skin cancer. This is why the existence of the ozone 'holes' in this layer, caused by chlorofluorocarbons (CFCs) previously found in such things as aerosol spray can propellants and air conditioning refrigerants, is of such concern. Fortunately, concerted effort by the countries of the world, which agreed in 1987 to the Montreal Protocol on harmful gases, means that CFCs have been phased out in favour of alternatives. There are encouraging signs that the ozone holes are becoming weaker, although they will still take some decades to recover fully.

**Above left** The ozone 'hole' in the southern hemisphere.

# Sun Safety

A great deal of harmful UV light is blocked by the protective ozone layer, but certain wavelengths still reach the Earth's surface. These are now known to be damaging, not only causing sunburn but also skin cancers. On the one hand, we are advised, rightly, to 'cover up' and limit our time in the sun, and to use sunblocking creams when exposed to strong sunlight. But, on the other, because we can't store vitamin D in our bodies for long we need to get out in the sun to 'top up' – some experts suggest around 5–30 minutes of exposure on bare skin twice a week should do it. It's probably best to avoid the hottest part of the day, around noon, or too much time on the beach, where sand reflects a lot of sunlight and the cooling sea breeze may be deceptive.

Covering up on the beach at Burton Bradstock, Dorset.

# Jet Streams

The jet streams that lie close to the top of the polar fronts are often mentioned in today's weather forecasts. They have a major effect on the weather at the surface as well as on how weather systems move and develop, especially in the middle la9titudes – such as in Britain. As we have seen, warm air expands and cold air contracts, and this has an effect on the exact pressure at any one height in the atmosphere. The altitude at which a particular pressure exists will be higher over a warm air column and lower above a cold one. This is especially important for the existence of the jet streams.

Meteorologists use charts (known as contour charts) that show the altitude of a particular pressure surface (often 300hPa). These charts indicate where there are warm and cold air columns, because a warm air column causes an upward bulge in the contours and, conversely, a cold column creates a depression. More significantly, the spacing of the 'contours' – the isobars – shows the path of winds at altitude. The closer the spacing of the isobars, the stronger the winds. Knowing this is helpful for seeing the path of the jet streams. From the surface, the most obvious visible sign of these winds is the motion of high cirrus clouds (see page 45).

In the northern hemisphere, there is a pool of cold air over the Arctic. This means there is a distinct difference between the heights of the air columns over that cold air and those over the warmer air to the south. Air pressure is higher at every altitude in the warm-air column, so a strong pressure gradient develops from south to north. Because of the Coriolis Effect this also produces extremely strong westerly winds at altitude. These winds form the polar front jet stream.

## What is a Jet Stream?

A jet stream is a ribbon of high-speed winds (moving at 90–108kph or more) that may be thousands of kilometres long, hundreds of kilometres wide, but just a few kilometres deep. All jet streams develop when there is a strong difference in temperature at different latitudes.

# The Jet Stream's Lobes

The northern polar front jet stream normally shows four or five lobes, with cold polar air extending south (in what are known as troughs) and warm tropical air extending north (in ridges). The troughs and ridges move around the Earth, following changes in the position of the jet stream.

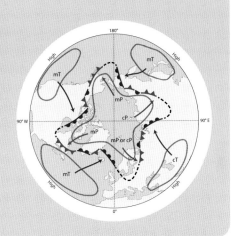

The polar front jet stream flows right around the world but, as you might expect, is not constant in its strength. It not only varies in speed but may even be absent in particular places, or split into two separate streams.

The temperature contrast is particularly great in winter, leading to stronger jet-stream winds at that time. Somewhat paradoxically, when the polar vortex is strong, the jet stream shows relatively small variations in latitude. But when the wind is weaker, the dips in latitude become much greater and 'polar lobes' of frigid air reach down towards the middle latitudes (and, correspondingly, intervening lobes of warm air move towards northern regions). In winter the main pool of frigid air lies over the Canadian Arctic.

When the polar vortex weakens, secondary pools of frigid air develop elsewhere, such as over the mid-Atlantic and over Siberia.

A somewhat similar situation exists in the southern hemisphere. Here, however, the large extent of ocean and the lack of large land masses produces a strong, more regular polar vortex. It is generally much stronger than its northern counterpart and tends to isolate the high atmosphere over the Antarctic from warmer air to the north. The strength of the southern polar vortex is one of the reasons why the southern ozone hole (see page 132) is much stronger than the intermittent ozone hole that is sometimes present in the north.

# The 'Greenhouse Effect' and Climate Change

The atmosphere as a whole is responsible for what has come to be called 'the Greenhouse Effect'. If the Earth had no atmosphere at all, the temperature would be around -18°C (-0.4°F), far too cold for life to exist. With its atmosphere, the Earth's average temperature is significantly higher, at about 14°C (57°F). This increase in temperature is caused by gases in the atmosphere (particularly carbon dioxide and water vapour) that act as a 'one-way blanket'. They allow radiation from the Sun to reach the surface but block some of the radiation that is reflected back from the ground at a different wavelength from escaping to space. The trapped radiation increases the Earth's overall temperature.

# Is the 'Greenhouse Effect' a good name?

The 'Greenhouse Effect' is actually not a very good analogy. When you enter a greenhouse, it does indeed feel warmer inside than it was outside. But this is because the Sun's radiation has warmed the whole interior, including the air, and the glass (or transparent plastic) has prevented the inside air from mixing freely with the cooler outside air. This is what causes the air to heat up and stay warm, rather than the transparent material directly blocking any outgoing radiation.

In the atmosphere itself, the various 'greenhouse gases', particularly carbon dioxide, have a major effect on the Earth's overall temperature. The increase in the concentration of carbon dioxide, mainly caused by the burning of fossil fuels, is undoubtedly the reason for global warming and climate change which, despite those who choose to deny this, are definitely happening and are likely to have far-reaching effects, such as a rise in sea level that will eventually flood many coastal areas and low-lying towns and cities.

**Left** Dramatic clouds over the English Channel.
**Below** Flooding at Lacock Abbey, Wiltshire.
**Pages 32–33** A cloudy day at Wimpole Hall, Cambridgeshire.

# 2. Clouds

Once you learn to recognise them, quite a few of the cloud types give useful hints on the current weather and on what changes may be coming. Many people are put off trying to learn which cloud is which: they seem to change from one to another so quickly that any attempt to identify them seems impossible. This indeed was the view for centuries, until Luke Howard, a British manufacturing pharmacist with a strong interest in meteorology and science generally, put forward his famous classification scheme in 1803.

# Cloud Classification

Clouds today are classified into ten main types (genera), which are easy enough to recognise and can also tell us a lot about the weather. Some of the individual species and varieties (in particular) are important for weather, and we'll get back to them later. As well as the main types, there are extra groups of supplementary features – but that's where most people give up in confusion!

The main types, or genera, can be subdivided in various ways. They fall into two main groups: cumuliform and stratiform clouds. Cumuliform clouds are known as 'heap' clouds ('cumulus' is Latin for 'heap'), whereas stratiform clouds are layer clouds ('stratus' is Latin for 'sheet'). The way clouds start to form is significant in reading how the weather might change – it is also significant to pressure systems and the ways in which they develop, but more

**Above** Cumulus clouds at Porthbeor beach, Cornwall.

# Howard's Cloud Types

Howard suggested that clouds could be classified in a way similar the way to how animals and plants are described, dividing them into genera, species and varieties. His scheme was so successful that some of his terms remain in use today, and you will probably recognise these four cloud types:

CIRRUS *fibrous or feathery clouds*

CUMULUS *heaped clouds with flat bases and rounded tops*

NIMBUS *rain-bearing clouds*

STRATUS *a level layer of cloud*

| CUMULIFORM CLOUDS | | STRATIFORM CLOUDS | |
|---|---|---|---|
| Cumulus | Cu | Stratus | St |
| Stratocumulus | Sc | Altostratus | As |
| Altocumulus | Ac | Nimbostratus | Ns |
| Cirrocumulus | Cc | Cirrostratus | Cs |
| Cumulonimbus | Cb | | |

on that in Chapter 3. All the cloud types are known by standard two-letter abbreviations.

An important type, cirrus, doesn't really fit into this two-group scheme, being composed of wisps of ice crystals. Cirrus is Latin for 'curl', an apt name for its elegant tendrils and curved shapes. It may neither be classified as a heap cloud nor as a layer. Sometimes all the ice crystal clouds: cirrus, cirrocumulus and cirrostratus, are instead grouped together as 'cirriform' clouds.

Another way of classifying clouds is by their altitude – by the atmospheric layer in which they occur – and this is perhaps the most convenient. There are three main layers within the troposphere and the ten genera of clouds can be described as to where they lie in these. Two types of deep cloud – the deep rain-bearing clouds – extend through two or more layers.

**Left** Dramatic clouds above Ickworth, Suffolk.

# Measuring Cloud Heights

**Although meteorologists normally use metric (SI – Système Internationale) units, cloud heights are always given in feet, because this is the form adopted years ago by the aviation industry.**

The clouds made from ice crystals – cirrus, cirrocumulus and cirrostratus – are in the highest layer (approximately 10,000–25,000ft [3,049–7,622m] in the polar regions and as high as approximately 20,000–60,000ft [6,098–18,293m] in the tropics). Because they are so high, the motion of cirrus clouds gives us clues as to the winds at upper levels. The clouds made up of water droplets – cumulus, stratocumulus and stratus – are in the layer nearest the ground (below about 6,500ft [1,982m]). As you might guess, in the middle layer (between 6,500 and 20,000ft [1,982 and 6,098m] or more, depending on latitude), the clouds tend to be a mixture of water droplets and ice crystals. These clouds are altocumulus and altostratus. Last but not least (and certainly the clouds we know only too well) are the deep rain clouds, nimbostratus and cumulonimbus. Nimbostratus generally occurs in the lowest and middle layers, but the largest clouds may reach even higher. Cumulonimbus clouds often extend through all three layers and even as high as the tropopause, just below the stratosphere, where they may spread out into flat-topped anvil clouds.

| LOW-LEVEL CLOUDS | | MIDDLE-LEVEL CLOUDS | |
|---|---|---|---|
| Cumulus | Cu | Altocumulus | Ac |
| Stratocumulus | Sc | Altostratus | As |
| Stratus | St | | |

| HIGH-LEVEL CLOUDS | | MULTI-LEVEL CLOUDS | |
|---|---|---|---|
| Cirrus | Ci | Nimbostratus | Ns |
| Cirrocumulus | Cc | Cumulonimbus | Cb |
| Cirrostratus | Cs | | |

# Cloud Types and Weather

Here is a description of the ten main cloud types, beginning with the low clouds. Later we'll look at some of the important sub-types (the species) that can help with predicting weather conditions (pages 57–65).

## Cumulus (Cu)

Cumulus clouds are the 'fair-weather' clouds – those small, fluffy heaps of cloud that children always draw. They form when the surface is heated by the Sun and invisible bubbles of warm air start to rise from the surface (the thermals, sought after by soaring birds and glider pilots). Cumulus clouds have flat, slightly darker bases. The bases of a group of cumulus clouds will lie at the same height above the ground (the condensation level), and their rounded tops are also at roughly the same level. The clouds are generally wider than they are tall. Cumulus clouds tend to appear with gentle or moderate winds and indicate that there will be no dramatic changes in the weather.

Some of the deeper forms (species) of cumulus provide some important clues on how the weather will develop (see pages 59 and 63–64).

**Below** Cumulus clouds over Erddig, Wrexham, North Wales.

# Stratocumulus (Sc)

Stratocumulus clouds are a layer of cumulus-type clouds spread out into a layer of stratocumulus cloudlets. There are always distinct breaks between the individual cloudlets but there may be considerable differences in their size. Sometimes the breaks are large enough to see blue sky between them. At other times you can't see any sky, but the breaks are given away by shafts of sunlight that penetrate through the layer. These 'crepuscular rays' are very common, although often mistaken for distant shafts of rain.

Stratocumulus is a very familiar cloud type – it is the most common type that forms over the oceans – and suggests that the weather situation is not changing rapidly. It often accompanies weak depression systems, when it may be quite thick, but it won't bring any rain to speak of, and perhaps none at all. Instead, stratocumulus clouds may produce fine drizzle or, if they have underlying cold air, a light powdering of snow or a fall of tiny ice crystals.

**Above** Crepuscular rays filtering through a layer of stratocumulus.

## Stratus (St)

Stratus is a featureless cloud that often shrouds the ground – especially high ground, such as hills or mountains, and even the tops of high buildings. To anyone within the cloud it appears as fog or mist, and so is a considerable hazard to walkers and road users. Stratus cloud often occurs behind the warm fronts of depressions. It also forms near the coast, when warm, humid air cools with the drop in temperature overnight. Low-lying fog that has formed overnight will often lift into a low layer of stratus as the morning sun begins to heat the ground or the wind rises. As with stratocumulus, stratus doesn't produce significant rainfall, but may give rise to slight drizzle, or a light fall of snow or minute ice crystals.

**Above** Low-lying stratus enshrouding Cat Bells and Causey Pike, seen from Derwentwater, Cumbria.

# Altocumulus (Ac)

Altocumulus is a middle-layer cloud that looks a little like stratocumulus, being also made up of small, individual cloudlets. But it is higher in the atmosphere, and it is normally easier to see blue sky between the individual elements. Unlike the cirrocumulus, which is even higher, altocumulus clouds always show shading.

Some species of altocumulus (described later) are important clues to forthcoming thundery weather, but generally they do not indicate any immediate, significant changes. Altocumulus often trail behind frontal systems, where the overall cloud cover is decreasing. They are also found ahead of approaching fronts where, if they gradually turn into altostratus, they are indicative of major changes to come.

**Above** Typical altocumulus clouds.

# Altostratus (As)

Altostratus is a relatively featureless, medium-level cloud. It is a good indicator of approaching persistent rain, particularly when it appears as an increasing and thickening layer. Altostratus generally follows cirrostratus and may initially appear with some darker streaks, but it soon changes to an almost uniform veil across the sky. The Sun, at first visible through the cloud, gradually disappears. As it does, it's as if the light is seen through ground glass, ceasing to cast any shadows. Altostratus itself rarely produces any rain that reaches the ground, but it precedes nimbostratus, which brings the long-lasting rain that does.

Altostratus is sometimes accompanied by ragged shreds of cloud beneath the main cloud layer. These are known as pannus (pan) and are a sign that the air is humid enough for the water vapour to condense. If you spot pannus, get out your umbrella – rain is on the way.

**Above** Altostratus at East Wittering Beach, West Sussex.
**Above right** Cirrus at Croome, Worcestershire.

# Cirrus (Ci)

Cirrus, one of the high cloud types, has a distinctive structure, being composed of ice crystals. It is always wispy or feathery in appearance. Although the crystals reflect light so that in the daytime they appear white and without shadows, occasionally cirrus may be so dense that it appears a darker grey.

Cirrus usually shows a denser head and long, trailing streamers of ice crystals (known as 'fallstreaks'). The heads tend to form in the region of strongest winds, and the way in which the fallstreaks trail behind them provides a clue as to the strength of lower winds. Fallstreaks can be hooked, curved, straight or appear quite random. When the trails are strongly hooked, this shows there is a major difference in the wind speeds, with the lower layer moving much more slowly. If, on the other hand, the fallstreaks are more or less vertical, this suggests that a deep layer of air is moving at a uniform speed – often a clue to approaching thundery weather.

How cirrus clouds move, compared with the motion of low-level clouds, is often an indication of how the weather is likely to develop. Bands of jet-stream cirrus (see page 72) are an important indicator of the approach and development of depression systems, and are the only way in which the jet streams are directly visible from the ground.

# Cirrostratus (Cs)

Cirrostratus is a thin veil of ice crystal cloud. It usually appears as cirrus cloud spreads out across the sky; this spreading is often the very first sign of an approaching depression. Initially cirrostratus is always very thin, and the Sun and Moon are visible through it. They are often accompanied by a halo, seen as a more or less complete circle around the Sun or Moon. This is caused through the interaction of light and the ice crystals that make up the cloud. The most common halo is white or shows just faint tinges of colour, and has a radius of 22°, which is approximately one hand-span at arm's length. If a halo appears it will only be visible for a short period, disappearing as the layer of cirrostratus becomes thicker. If you see a fairly well-defined, narrow but essentially colourless halo, it has almost certainly been produced by cirrostratus cloud.

**Above** Cirrostratus with a halo.

**Above right** Cirrocumulus near Totnes, Devon.

# Cirrocumulus (Cs)

The last of the three high clouds is cirrocumulus. As the name suggests, this is a layer of tiny cloudlets, again consisting of ice crystals, which is often difficult to distinguish from high altocumulus. Technically, it is defined by the size of the individual cloudlets, which are smaller than those of altocumulus and do not show any signs of shading. It is relatively rare, and also offers no particular signs of forthcoming changes to the weather.

# Nimbostratus (Ns)

Nimbostratus is a deep, dark grey layer of cloud and is the main rain-bearing cloud in depressions in both summer and winter. If altostratus thickens and lowers, and the rain starts to fall, the cloud has turned into nimbostratus. The rain will be more or less continuous and may persist for many hours.

Nimbostratus may be very deep and often extends down almost to the surface, frequently accompanied by ragged shreds of pannus beneath the main cloud layer. In winter, if the rain is falling into a layer of cold air below freezing, nimbostratus may produce long periods of snowfall.

**Above** Nimbostratus at Littondale, North Yorkshire.

**Right** Cumulonimbus at Teesdale, County Durham.

## Cumulonimbus (Cb)

The other rain-bearing cloud, cumulonimbus, is related to cumulus and often begins in a similar fashion. But instead of layers of fluffy clouds, the multiple individual rising cloud cells combine into one big cloud mass, often huge and very impressive. Frequently the domed tops of major developing cells are distinctly visible in the growing cloud. Cumulonimbus clouds produce showers and form thunderstorms. The clouds are very deep, and often extend throughout the whole troposphere. They are only limited in their upward growth by the inversion at the tropopause. Because the tops of most cumulonimbus clouds rise so high in the atmosphere, they reach the freezing level, so the cloud there actually consists of ice crystals, creating a top of false cirrus. The top may spread and flatten in that characteristic anvil shape. Anvil or incus (the Latin for 'anvil') is one of those supplementary features that help to identify particular clouds. Clouds in this particular form are known as cumulonimbus incus (Cb inc).

The base of this cloud form is very dark and may look somewhat like heavy nimbostratus. However, unlike the heavy, sustained rain from nimbostratus, the rain from cumulonimbus clouds tends to be intense but dies away after perhaps 30 minutes or so. However, just as you're putting away your umbrella, another cloud may well follow soon after, with its own accompanying burst of rain.

## Mamma

Another supplementary feature often associated with cumulonimbus clouds is called mamma (Latin for 'breasts'). These pouches of cloud often hang below an extensive anvil and may look very dramatic, especially when illuminated by a low sun. For once, they are a good weather sign, generally appearing as a cumulonimbus cloud retreats, and are a clue that the worst of the shower is over.

**Below** Mamma seen under a cumulonimbus cloud.

# How Clouds Form

Because clouds are so important both as the source of particular weather conditions and for predicting the weather to come, it's useful to know a bit more about how they are organised, and how they develop.

**Below** A glorious cloudy sky above Box Hill, Surrey.

## CLOUD FORMATION
Clouds form in three ways.

**1.** By heating from below, known as convection.

**2.** By air being forced to rise and cool, for example when the wind forces air to rise over hills or mountains. This is known as orographic lifting.

**3.** When warm air is undercut by cold air at the boundary between air masses, and the warm air is lifted. Because the boundary between air masses is called a front, this is known as frontal lifting.

These three methods of cloud formation are matched by three different ways in which rainfall is produced – this aspect is discussed in Chapter 3.

For cloud formation by convection, the ground is heated by the Sun, and convection causes invisible bubbles of warm air to rise from the surface. The same thing can happen when cool air passes over warm water. While these individual rising parcels of air – cells, or thermals – continue to be warmer than those around them, they will keep ascending. Cumulus and cumulonimbus are produced in this way.

Sometimes the wind will carry these clouds over higher ground, and then the clouds will grow even deeper, as a result of the orographic lifting. Cumulonimbus, in particular, may grow so much that they produce showers of rain. Orographic lifting is also involved when warm, moist air is carried over high ground and a layer of stratus often forms over the hilltops. Anyone on the hills becomes shrouded in mist or fog. The surface need not be particularly high for this to happen; it can even happen when a stream of warm air over the sea encounters a coastline.

Finally, frontal lifting is a method of forming clouds found in depression systems, and we'll get to these clouds in more detail when depression systems are discussed in Chapter 3.

As an example, cumulus clouds arise when the ground is heated by the Sun, which produces rising thermals. As you may remember, when air rises, the pressure drops, and the air expands and becomes cooler. The decrease in temperature of the rising thermal occurs at a specific rate, known as the lapse rate. This rate is independent of the surrounding air, as long as the latter is still cooler than the thermal. The thermal air parcel will continue to rise until it reaches the same temperature as its surroundings.

Eventually the parcel of air may, still rising, become cool enough for the moisture within it to condense into tiny cloud particles. At that point the air has reached its dewpoint – it is condensing more than it is evaporating. After condensation has set in within a cloud, what happens next depends on the exact conditions. When condensation (or freezing) takes place, heat – known as latent heat – is released. This warms the parcel of air and helps it to continue to rise, albeit at a slightly different rate.

# Condensation and Freezing

Water vapour in air cannot condense in isolation. There needs to be minute particles, such as dust particles derived from soil or salt crystals originating in the sea, to act as 'condensation nuclei'. In reality, these particles are present in such large quantities throughout the atmosphere (even over dry deserts) that condensation readily occurs anywhere in the world.

After condensation comes freezing (known as glaciation to meteorologists), where conditions are much more critical. For freezing to happen there must be either a solid surface or freezing nuclei that are of a particular shape, often derived from clays. These special nuclei may be relatively rare. If there are no freezing nuclei, water droplets may remain liquid at temperatures well below 0°C (32°F) and in fact won't freeze spontaneously until the temperature has dropped to -40°C (-40°F).

These non-frozen 'below freezing' droplets are said to be 'supercooled' – actually, such conditions are surprisingly common in the atmosphere. Some high cloud forms, such as altostratus and altocumulus, are classed as 'mixed' clouds. This is because they often consist of a rather chilly mixture of unfrozen, supercooled droplets and actual ice crystals.

**Below** Icicles in Snowdonia National Park, Gwynedd, North Wales.

# Aircraft Icing

Supercooled droplets freeze instantly if they come into contact with a solid surface. This is why aircraft flying into clouds containing supercooled droplets are at risk of 'icing' – when the wings and other surfaces rapidly become coated in solid ice.

# Black Ice

Raindrops sometimes exist in a supercooled state, so that when they fall into a layer of extremely cold air next to the Earth's surface, they will freeze as a clear glaze on hitting the ground or objects on it. This is 'black ice' and may even happen when the air temperature is very slightly above 0°C (32°F).

**Below** Treacherous black ice in Carmarthenshire, South Wales.

# Ice Storms

Glaze often forms along an approaching warm front. If the air ahead of the front is below freezing, the falling rain freezes immediately on contact with any objects. This is the form of ice that occasionally produces dramatic 'ice storms'. These can appear beautiful, coating winter tree branches with a magical, icy appearance. However, they are also highly destructive – a major ice storm that hit Canada and the north-eastern United States in 1998 broke down trees, telephone and power lines, and caused electricity pylons to collapse.

# Stable and Unstable Air

While conditions are such that an air parcel continues to rise, the air is said to be unstable. If the parcel of air cools to the same temperature as its surroundings (regardless of whether condensation has set in or not) it must come to a halt. Such air is said to be stable. If any process – such as orographic lifting over a hill – forces it to rise further it will become colder than its surroundings, and attempt to sink back down again.

The different conditions that produce stability or instability lead to the formation of different types of clouds. Cumulus, for instance, are unstable clouds in which the air is unstable.

# Cumulus Species

The ten main varieties of cloud described in the last section provided some important clues to the weather, but some of the individual forms (species) within them are also important for understanding developing weather conditions. Species are indicated by three-letter abbreviations, which follow the two-letter abbreviations for the main cloud genera.

## Cumulus Fractus (Cu fra)

Cumulus start out as ragged wisps of cloud, called cumulus fractus. In the early morning these may be little more then misty patches, which may decay and not even make it to becoming clouds. If they are produced by rising thermals they may show a slightly domed top, but even then these ragged bits of cloud might not get any further. Towards the end of the day, when heating by the Sun has decreased, cumulus may dissolve into ragged fragments of cumulus fractus cloud. These little cloud fragments don't tell us a great deal about the weather.

# Cumulus Humilis (Cu hum)

The simplest cumulus clouds are known as cumulus humilis. Here the clouds appear flattened, and are shallow, so have a far greater horizontal extent than vertical depth. This flattening happens for either of a couple of reasons: they may start off early in the day, but then grow vertically – eventually becoming another cumulus species, or it may be that conditions are suppressing their upward growth – this is a sign that warm air is starting to move in above them, and so indicates a depression system may be on the way.

**Above** Cumulus humilis over East Head, West Sussex.

## Cumulus Mediocris (Cu med)

Cumulus clouds that continue to grow turn into the next, larger species: cumulus mediocris. The individual clouds are deeper, and they are about as deep as they are wide. They have domed tops and they are likely to grow upwards further.

# Cumulus Congestus (Cu con)

If growth continues upwards, these clouds may become cumulus congestus. Here, the towering clouds are much deeper than they are wide and may show signs of vigorous upward growth. They look very impressive and as if they might bring a lot of rain. But in fact, under normal circumstances, for most of the year, no cumulus clouds produce rain.

Just occasionally in summer, over Western Europe and Britain, a cumulus congestus may become so deep that it is able to create raindrops through the collision of vast numbers of the individual, microscopic, cloud particles within it. This process is known to meteorologists as 'warm rain' (see page 94), but that has nothing to do with the actual temperature of the rain.

Cumulus clouds are confined to the lowest layer of the atmosphere, below 6,500ft (1,982m). Sometimes an inversion (an increase of temperature with increasing altitude) creates a stable layer and prevents cumulus from growing upwards. The thermals may still reach the dewpoint and condensation occurs, but they are prevented from rising even further. When this happens, the clouds spread out sideways and may give rise to a layer of stratocumulus. A similar process may occur at higher altitudes to produce a layer of altocumulus.

**Below** Cumulus mediocris over the sea.

**Below** Cumulus congestus.

# Cloud Streets

Cumulus clouds are often organised into long, more or less parallel lines of individual clouds. These are known as cloud streets and are particularly striking in many satellite images, especially where cold air is streaming off the land and out over the warm sea. When they are over land, cloud streets may be linked to specific, persistent sources of thermals, producing a succession of individual cumulus clouds that are carried downwind. We still don't really know the exact mechanism leading to the strikingly regular spacing of these lines of cloud.

**Above** Cloud streets over Pentire Head, Cornwall.

# Altocumulus Species

As I mentioned earlier, altocumulus clouds don't usually point to dramatic changes in the weather, but two species of middle-level cloud are strongly associated with thundery conditions. These are altocumulus floccus and altocumulus castellanus. Both are strong indications of instability at cloud level.

## Altocumulus Floccus (Ac flo)

Altocumulus floccus consists of isolated tufts of cloud, generally white or grey. 'Floccus' comes from the Latin for a 'lock of wool', rather than any reference to flocks of clouds. They are disorganised in appearance, although they do tend to form in lines along the upper-level wind.

**Above** Altocumulus floccus.

## Altocumulus Castellanus (Ac cas)

Altocumulus castellanus (from the Latin for 'castle'), show distinct 'turrets' of cloud, billowing upwards from the general layer. Again, such clouds often form as lines of cloud, and are often associated with moderate turbulence and even icing conditions. These two species of altocumulus indicate that there is instability in the middle level of the atmosphere. They are harbingers of heavy showers and thundery conditions. If cumulus clouds grow deep enough to reach up to this layer, they gain additional energy from them to grow even further. Altocumulus clouds may help to produce cumulonimbus clouds, which may reach as high as the tropopause.

**Above** Altocumulus Castellanus in Devon.

# Cumulonimbus Species

Cumulonimbus clouds, as discussed in the previous chapter, commonly produce relatively short-lived showers of heavy rain. Sometimes they develop into much stronger systems, as you will find out in Chapter 4.

The two species of cumulonimbus provide an indication of the processes taking place in the top of the growing cloud, important in predicting its future behaviour. Both species are unique to cumulonimbus, so if either is visible in the cloud tops, you can be sure the cloud concerned is a cumulonimbus. The species mark the transition from a cumulus congestus cloud to a cumulonimbus. Strange as this may sound, examining the tops of clouds through binoculars is a good way to see exactly what's going on as these clouds develop.

## Cumulonimbus Calvus (Cb cal)

The Latin word 'calvus' means 'bald', but this gives the wrong impression, since this cloud species has quite a lot on top. But rather than the normal, hard and smooth outline – the classic 'cauliflower' cloud shape that is visible at the top of most rising cumulus cells – the outline becomes softer. This change occurs when the water droplets in the cloud start to freeze into ice crystals.

This is an important stage in the development of the cloud, indicating the start of the processes that are going to lead to heavy rain and hail, but it is often rather difficult to detect.

**Below** Cumulonimbus Calvus.

# Cumulonimbus Capillatus (Cb cap)

As freezing (glaciation) continues in the tops of cloud cells, the cumulonimbus calvus stage passes and the cloud-top assumes an obviously fibrous appearance, known as the species capillatus. In effect, the top of the cloud turns into cirrus. The exact appearance will depend greatly on the winds at that level. When the wind is gentle, the ice crystals fall more or less vertically, and the resulting trails (fallstreaks) produce strong vertical striations to the top of the cloud. When winds are stronger, the crystals spread out and may create a large, overhanging anvil of false cirrus. Occasionally, if the convection within the cumulonimbus cloud is particularly strong, the cloud-top may reach the jet stream, and the ice crystals may form a vast cirrus plume that is rapidly carried downwind. The top of the cumulonimbus seems to 'explode' and rapidly cover a large area of the sky.

**Below** Cumulonimbus capillatus.

# Wave Clouds

One form of orographic lifting may result in distinctive cloud forms, called wave clouds. Air flowing over obstructions, such as a range of hills, is set into a series of waves behind the high ground. If there is a layer of stable, humid air at height above the ground, the forced upward motion may lead to condensation in the tops of the waves, creating a series of clouds above and downwind of the obstruction, with clear gaps between them. The clouds are smooth and often lenticular (lens-shaped) – the 'flying saucer' look of a sci-fi UFO. They may appear behind a single hill but are most striking when the wind is approximately at right-angles to a range of hills. Sometimes there are no clear gaps between separate clouds. Instead, there is a streamer of unbroken cloud downwind of the obstruction, and the wave nature is given away by the rise and fall of its base and top.

Unlike most clouds, lenticular (or wave) clouds remain stationary and do not drift downwind. They will remain in the sky while the wind direction and strength remain constant. If you examine them carefully, perhaps through binoculars, you can see that individual clouds may be forming on the upwind side and decaying downwind. These clouds are known as the species lenticularis (abbreviated to len). The most commonly seen are altocumulus lenticularis (Ac len), although lower, stratocumulus lenticularis (Sc len) and higher cirrocumulus lenticularis (Cc len) are also fairly common.

**Above** Lenticular clouds in North Yorkshire.
**Pages 66–67** A rainy day over Langdale Pikes, Cumbria.

# 3. Highs and Lows

The majority of changeable weather results from the succession of low-pressure areas (depressions) and high-pressure anticyclones that pass across the country. It helps considerably to understand how these systems occur and to recognise the signs of change.

# Depressions

A depression (also known as a 'low') is a low-pressure system that occurs in the mid-latitudes. It is also referred to as a mid-latitude cyclone – there are other types of cyclone that occur elsewhere. Depressions are responsible for the most changeable weather in the middle latitudes, including the British Isles. Understanding how depressions develop and evolve can help to understand how weather is likely to change over time. This chapter explains how mid-latitude depressions work in the northern hemisphere, where the surface winds flow anticlockwise into the centre, and the air flows out, clockwise, at height.

## Here Comes the Rain

When TV weather forecasters warn us of an incoming low-pressure system – or depression –this usually means unsettled weather is on the way. Where warm air flows into a low-pressure centre, the air cannot accumulate there indefinitely and eventually rises, causing its pressure to decrease. The air expands and cools and produces extensive precipitation – rain or snow.

# The Polar Front

Depressions occur at the polar front, the boundary between cold polar air from the polar cell and the warm, tropical air from the lower-latitude Ferrel cell. Typically the air masses involved are maritime polar (mP) and maritime tropical (mT) air. In reality, frontal boundaries are not the sharply defined lines familiar from some weather charts, but an area or 'zone', around 100–200km (62.1–124 miles) in width, at the surface. The jet stream, as mentioned on page 28, is a band of strong wind that meanders as it circles the globe, sits just on top of the polar front and drives this unstable area's behaviour.

The polar front boundary may initially be roughly straight and unwavering, with easterly and westerly winds on either side of it. But it is unstable, and will begin to distort into a shallow wave, rather like a wave propagating in water, which tends to move eastwards. These secondary waves may go on to develop into fully-fledged depressions, as described here, or form lesser systems that may also bring bad weather (we'll get to those later).

This wave has distinct warm and cold fronts, both of which bring unsettled weather. At the warm front, the warm air slides up over the cold air that lies at the surface. Following the warm front is the centre of the low-pressure depression (weather can actually clear up for a while as this passes over) and then the cold front, which may trail a long way behind the centre of the depression.

The warm (tropical) air is itself undercut by the following cold (polar) air, which progressively gains on the warm air. The transition where cold air replaces the warm air forms the depression's cold front.

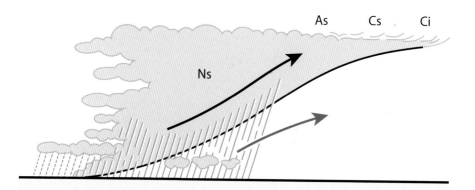

# Warm-sector Stage

As the wave distortion in the boundary increases in size, a warm sector of air, triangular in shape, with its apex at the centre of the depression, develops between the warm front on the east and the cold front on the west. The winds at both the warm and cold fronts may suddenly change direction and on the weather chart the isobars will show a sharp 'V'-shape at both the warm and cold fronts indicating this. At this warm-sector stage the cloud cover has increased, as has the area where it is raining.

The jet stream, which lies above the low-pressure centre near the point where the cold and warm fronts meet, has developed a distinct curve, forming an 'S' shape overlying the surface system. As the depression develops, the curvature of the jet stream becomes even more pronounced. It flows roughly north-east behind the cold front, turns sharply over the low-pressure centre and flows south-east ahead of, and roughly parallel to, the warm front.

Ahead of the warm front the upper wind will be from west to east, possibly north-west to south-east, and the lower wind comes from a southerly direction – indicating that a depression is off to the west. But if the upper wind is from a southerly direction, say south-westerly, and the lower wind is from the north or north-west, you are likely positioned behind the cold front – the depression centre lies to the east. When the upper and lower winds are roughly parallel to one another, above the warm sector, or in opposite directions, to the north of the pressure centre, the weather is unlikely to change rapidly.

At this stage the whole depression moves east, roughly parallel to the isobars in the warm sector. The cold front moves faster than the warm front and gradually reduces the size of the warm sector. It eventually catches up with the warm front completely, eventually pushing the warm air upwards off the ground, creating what's known as an occluded front. So now there are three air masses with

**Below** A warm-sector depression.

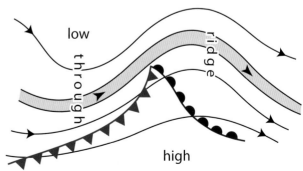

different temperatures in contact with one another and three distinct fronts: warm, cold, and occluded. The surface pressure chart (weather charts showing areas of high and low pressure) indicates a point, known as the triple point, where the three fronts meet. This point is normally under the jet stream. During this the cloud cover has increased yet more and now blankets a large area, particularly if the occluded front develops into a large spiral. Depending on the motion of the whole system, occluded fronts may bring extended periods of heavy cloud and rain.

## Late-stage Depression

By the time an occlusion starts to form, the depression is generally deepening and intensifying and the strongest winds develop around its centre. But occlusion normally signals that the depression will soon start to decay. Its centre moves away from the triple point, generally in a northerly direction, sometimes also moving westwards. It may trail a long occluded front behind it (see below), sometimes resulting in an extended spiral of cloud and rain.

By now, practically the whole of the warm sector has been undercut and the system starts to 'fill', with the central pressure increasing. In most systems, the polar front has pushed further south behind the depression and the next secondary wave and incipient depression forms at a lower latitude. Occasionally a new secondary depression develops rapidly and, as it moves east, the old, decaying depression to the north has a tendency to move westwards – the two low-pressure centres 'circle' around each other.

During the decaying stage, the warm air gradually cools and tends to descend towards the surface. The cloud cover reduces and the fronts and the cloud associated with them also changes, which we will discuss later.

**Below** A late-stage depression.

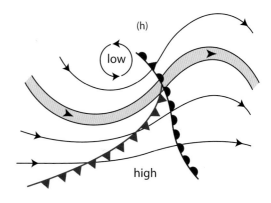

# An Approaching Warm Front

Now we have looked at how active depressions form and develop, let's learn more about the movements of the warm and cold fronts and how they may be observed.

The first signs of an approaching depression are often found in the high jet-stream cirrus clouds. As mentioned earlier, the direction of the jet stream can often be seen in the motion of high-level cirrus clouds. When the direction of the jet stream is sharply in contrast to winds at the Earth's surface – where the winds are 'crossed' – changeable weather may be expected.

The cirrus streams in a generally easterly direction, approximately parallel to the line of the approaching warm front. When the wind at height is 'crossed' with the wind at low level this can be observed from the motion of cumulus clouds. As a wedge of warm air moves in above the cold, it tends to subdue air rising by convection. This is because the upper cloud cover increases, and that lessens the amount of sunlight heating the ground below. The warm air may create an inversion, which blocks the upward growth of clouds. Fair-weather cumulus that may have been present originally now become flattened (cumulus humilis).

Although the exact height of any cirrus clouds will vary, a typical height might be 30,000ft: nearly 10km. (This is also a normal altitude for high-flying aircraft.) The slope of warm fronts is very shallow: about 1:100 or 1:150. This means that the very beginning of the wedge of warm air moving in aloft may be some 1000–1500km (621–932 miles) ahead of the surface front.

**Above** Croome, Worcestershire.

# An Arriving Warm Front

The weather is definitely on the move, but how long will it take for that warm front to arrive? This is hard to tell, even for professional forecasters. The changes in the cloud cover and the strength and direction of the wind are useful signs.

The time until the warm front arrives depends on so many factors, including whether the central pressure of the depression is dropping. A typical speed of advance is about 50km/h (31mph), so the rain may be expected some 20 hours after the very first high cirrus is seen. As another clue, the Sun usually disappears behind thickening altostratus about halfway through the sequence. The band of rain along the warm front will take about 5 or 6 hours to pass.

High cirrus is not always visible, but you can also get hints of what's to come from reading the condensation trails of high-flying aircraft. When the trails spread out sideways and are very slow to fade, that's an indication that warm, humid air is spreading at that altitude.

Watch out for high cirrus streaks gradually becoming more numerous, often spreading out into a general veil of cirrostratus cloud. This is when a solar halo (see page 153) often appears, although perhaps visible only for a short time. The cirrostratus becomes denser, and the cloud layer steadily lowers,

eventually turning into altostratus. Initially, this may show some streaks, before it slowly thickens too. The Sun appears as if seen through ground glass and ceases to cast any shadows. Eventually, its location is no longer visible. Any slight precipitation from altostratus won't reach the ground. Occasionally, there may be ragged shreds of cloud (pannus, see page 44) visible below it.

The altostratus continues to thicken and its base descends even closer to the ground. When the rain arrives, it is safe to say that the altostratus has turned into nimbostratus. This very deep, dark cloud, often accompanied by low-level pannus, often just precedes the arrival of rain (or snow) at the ground. How long the rain lasts will depend on where you are in relation to the centre of the depression. Well away from the centre of the depression, it may pass relatively quickly, but close to the centre it may persist for several hours. Although the rain may seem continuous and unbroken, in general there are bands of heavier rain running parallel to the front, with regions of lighter (or no) rain between them.

Another obvious clue to an approaching warm front is the way the wind changes: it backs (see page 74), becoming more southerly, and increases.

## Backing and Veering Winds

• A backing wind is one that changes direction anticlockwise, for example from south-east to north-east.

• A veering wind is one that changes direction clockwise, for example from south-west to west.

As the front approaches the pressure also drops at an increasing rate and the air temperature decreases. Depressions are also accompanied by changes in humidity (specifically, the dewpoint at which condensation occurs) – it's possible to detect them with even simple meteorological equipment, such as the weather stations widely available nowadays.

## SIGNS OF AN APPROACHING WARM FRONT

| | |
|---|---|
| **Cloud cover** | Cumulus become flattened. High cloud encroaches. Overall cloud increases in the sequence: cirrus, cirrostratus, altostratus, nimbostratus |
| **Wind** | Increases and towards south. High- and low-level winds may be crossed |
| **Pressure** | Drops at an increasing rate |
| **Temperature** | Steadily decreases |
| **Visibility** | Initially good, but decreases |
| **Dewpoint** | Remains steady before the front passes |
| **Precipitation** | Initially none, but virga and pannus may appear beneath altostratus. Finally rain (or snow) arrives with nimbostratus and is more or less continuous |

# A Departing Depression

What weather changes occur as a depression passes will depend on where you are standing in relation to its centre. As an example, imagine the depression's centre passing north of you, and in doing so the warm sector passes right over where you are. When the warm front arrives it will take some time to pass over, because it is a zone some 100–200km (62–124 miles) wide where there is mixing between the two air masses. As it goes by there will be some distinct changes in the weather conditions.

| WEATHER CHANGES AS A DEPRESSION PASSES | |
|---|---|
| **Cloud cover** | Normally remains nimbostratus, but may alter to dense stratocumulus or heavy stratus |
| **Wind** | The wind veers from southerly (south-east or south) to south-west |
| **Pressure** | Steadies, ceasing to fall |
| **Temperature** | Cool, because of the downdraughts caused by the rain |
| **Visibility** | Generally deteriorates |
| **Dewpoint** | Rises with the arrival of warm air |
| **Precipitation** | Generally ceases or changes to light drizzle |

# Conditions in the Warm Sector

Within the warm sector, the direction and the strength of the wind remain steady, with the direction being generally south-westerly. The pressure also remains steady, unless the depression is actively deepening. The cloud cover in the warm sector area depends greatly on the distance from the centre of the depression. Close to the centre, the cloud may be more or less continuous stratocumulus or stratus, but further away it tends to more broken. Sometimes the warm air is unstable, and cumulus clouds may develop. There may also be remnants of the higher cloud that accompanied the warm front, and this may produce skies of mixed cloud. If there is any precipitation, it will be light.

| WARM SECTOR CONDITIONS | |
|---|---|
| Cloud cover | Stratocumulus or stratus close to centre, more broken further away, even with new cumulus |
| Wind | Steady, generally south-westerly |
| Pressure | Remains steady |
| Temperature | Tends to rise |
| Visibility | Often poor, fog is sometimes present, especially over the sea |
| Dewpoint | Remains steady |
| Precipitation | Occasionally light rain or drizzle close to centre |

# A Cold Front

The approach of the cold front, in the warm sector, is far more difficult to observe than it was for the warm front. The low cloud cover (especially close to the centre of the depression) usually prevents a clear view of the sky. With an active cold front (where warm air is rising) the sequence of clouds is actually the reverse of that seen at the warm front. But as the slope of a cold front's wedge of air is much steeper (at about 1:75) than that of a warm front, the changes that occur as it passes are more rapid. The rear of the cold front may be quite sharp, with the rear of the cloud mass lying in a straight line across the sky.

Although the wind generally remains steady within the warm sector, it may increase and even back to some extent just ahead of the cold front. At the same time, the pressure may drop slightly. When the cold front arrives, however, it brings quite dramatic changes. The wind veers strongly, often changing from south-westerly to westerly or even north-westerly. This change is often accompanied by severe squalls. The temperature drops suddenly and the pressure rises rapidly. The changes in wind direction and temperature, together with the arrival of clear air with good visibility, are the most noticeable signs of the cold front's arrival for the majority of observers.

Although most of the heavy rain may fall from nimbostratus cloud, some convective clouds (cumulonimbus) are often mixed in with the layer cloud. These produce sharp pulses of rain, sometimes accompanied by hail or thunder. When the cold front has passed and the rear of the front becomes visible, the tops of individual cumulonimbus clouds (sometimes with distinct anvils) are often seen, embedded in the general line of cloud.

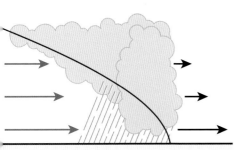

**Above** A typical cold front.

## THE CHANGES AT A COLD FRONT

| | |
|---|---|
| **Cloud cover** | Rain-bearing nimbostratus may have some embedded cumulonimbus |
| **Wind** | Backs sharply to westerly or north-westerly, often with sudden squalls |
| **Pressure** | Rises suddenly |
| **Temperature** | Drops abruptly |
| **Visibility** | Poor in the rain |
| **Dewpoint** | Suddenly drops with the arrival of cool air |
| **Precipitation** | Heavy rain, sometimes accompanied by hail or thunder |

# Weather Behind the Cold Front

With the arrival of the cold maritime polar air behind the cold front, weather conditions also change dramatically. Immediately behind the cold front the sky may be clear, but cumulus and cumulonimbus clouds – the latter with accompanying showers – start to develop. Satellite images usually show the area behind the front as speckled with shower clouds that have built up in this unstable air.

As before, if the upper and lower winds are 'crossed', this is an indication of changeable conditions. In this case the jet stream is roughly parallel to the cold front, and thus generally south-westerly, whereas the lower wind may be from the north-west. In this situation, there are likely to be clear skies, interrupted by heavy showers.

**Below** A long, trailing cold front and short occluded front north of the depression centre.

| CHANGES AS THE DEPRESSION MOVES EASTWARDS | |
|---|---|
| **Cloud cover** | Clear immediately behind the cold front, then cumulus, cumulonimbus and showers |
| **Wind** | May back slightly, becomes gusty and stronger |
| **Pressure** | Rate of rise decreases slowly |
| **Temperature** | Steadies |
| **Visibility** | Excellent |
| **Dewpoint** | Remains steady |
| **Precipitation** | Often none immediately behind cold front, then individual showers |

# When the Low Centre Passes South of the Observer

The changes described on page 75 relate to a depression passing north of the observer. If you are observing a depression passing south of you, the situation is less clear-cut. Instead of any distinct changes as the warm front, the warm sector and the cold front pass, there will be a gradual increase in cloud cover and possibly a prolonged period of rain. Jet-stream cirrus may give a warning of weather changes to come. The changes in the cloud cover will be similar, but the overall amount of precipitation will normally be less, and the convective activity found at a cold front is absent. The wind tends to back continuously from south-easterly ahead of the depression, through easterly to northerly and even north-easterly.

If the depression has reached the stage at which the warm air has been undercut by the cold air mass, then an occluded front will have developed. In a cold occlusion the air mass following the front is coldest, whereas in a warm occlusion the coldest air lies ahead of the front. Both types produce a similar succession of clouds as those at a warm front, followed immediately by the showers and lessening rain that accompanies a cold front.

## An Occluded Front

Generally, an occlusion starts to form as the depression is deepening and intensifying. It is often at this stage the strongest winds develop around the centre. But after occlusion has started, the depression generally begins to decay. The centre of the low moves away from the triple point and may trail a long occluded front behind it, sometimes resulting in a long spiral of cloud and rain. Although the temperature contrasts are decreasing and the occluded front is weakening, if the motion of the depression carries that front over the observer, there may be a very prolonged period of rain.

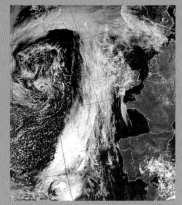

**Above** A long occluded front, wrapping around the depression centre.

# Subdued Frontal Systems

What about the situation we mentioned earlier, in relation to decaying depressions, especially where the warm air is descending? There are many entire low-pressure systems of this sort, where the weather is subdued in comparison with active depressions. In these weak systems the warm air is descending at both warm and cold fronts, which are known as 'kata fronts' (from the Greek word for 'down'). This subsiding air is particularly stable and the usual sequence of cloud formation at warm fronts is absent. Both behind and ahead of the warm front, the subsiding air limits the growth of clouds. Ahead of the warm front, the inversion caused by the subsiding air in the middle troposphere limits the growth of low-level cumulus. The cumulus gradually thickens into heavy stratocumulus, both ahead of and at the warm front. This stratocumulus persists across the warm sector, often with some fairly heavy stratus cloud. The cloud is often warmer than 0°C (32°F), especially in summer, so any precipitation is in the form of light rain or drizzle.

The stratocumulus and stratus continue and thicken at the kata cold front, still with little precipitation. Behind the front, the layer clouds eventually clear, to be replaced by cold unstable air, as in an active system, with cumulus and cumulonimbus clouds.

## Stationary Systems

Sometimes both active and subdued systems may come to a halt. This usually brings a prolonged period of heavy, overcast cloud and, often, persistent rain. Accompanying this, troughs of low pressure tend to swing (anticlockwise) round the stationary low centre. These troughs may bring pulses of rain. As with more pronounced fronts, a clue to their arrival is when the wind backs slightly. They may also bring a thinning of the cloud layer, which appears to brighten up, although no true breaks appear.

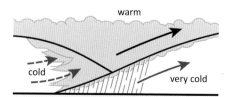

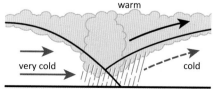

# Cold-front Waves

As mentioned previously, all depressions start as minor waves (secondary waves, also known as cold-front waves) on the polar front. Such waves commonly develop on the trailing cold front from a major depression, but do not always develop into fully-fledged depressions themselves, but they do bring a period of rain. Often as the main cloud and rain of a major depression is clearing away, there may be a lightening of the skies, the rain ceases, and you may hope that clearer weather is approaching. The sky remains cloudy, however, because a cold-front wave may be following immediately behind the receding depression. The cloud increases and another period of rain begins. The pattern of winds around such waves tends to be unpredictable, with air flowing into the tip of the wave. However, as the wave passes, a true clearance finally takes place.

# Thermal and Polar Lows

During the summer, the Sun heating the land may be enough to produce a low-pressure area. Sometimes it will be strong enough to create an area with a closed isobars and circulation system of its own (a thermal low), but it is more likely to distort the overall pattern of isobars and produce what is called a thermal trough. Both forms tend to die away at night, once the sun goes down and heating ceases, but they may generate sufficient instability during the day to create cumulonimbus clouds and the resulting showers, or even thunderstorms.

Another form of low may happen when frigid arctic or cold polar air flows over a relatively warm sea, with most effect in winter when there is the largest contrast in the air and water temperatures. Unlike the summer thermal lows, reliant on daytime sun, the water heats from below continuously, and these polar depressions (or polar lows) may become very intense, with a closed circulation.

**Right** Sun shining through trees at Petworth, West Sussex.

# High-pressure Systems and Depressions

We tend to associate high pressure with fine, sunny weather. If the forecaster tells us that the Azores High is extending a ridge of high pressure over the British Isles, we assume we're in for a settled sunny period. But, sadly, this is not always the case. Most changeable weather in the middle latitudes accompanies depressions, but the air that rises in them must come from somewhere – and it flows in from neighbouring high-pressure areas (anticyclones). In the northern hemisphere, anticyclones, like depressions, also move from west to east under the influence of the general westerly flow.

# Conveyer Belts

There are three distinct flows of incoming air associated with a depression, known as 'conveyor belts'.

The cold conveyer belt runs approximately towards the north-west of a depression and lies ahead of the warm front. It feeds cold air into the depression system, coming from an anticyclone (a high pressure centre) to its east. This flow of incoming air gradually rises, then turns and flows back towards the east at middle altitudes. The warm conveyor belt flows parallel to the cold front, rising as it does so. It brings warm, humid air into the system and is fed by another anticyclone that lies to the south of the depression. It eventually also turns towards the west and its warm air overlies the cold conveyor belt. This is the flow of air that brings most of the cloud and rain (or snow) to the depression. A third conveyer belt of cold, dry air runs, generally towards the north-east, behind the cold front.

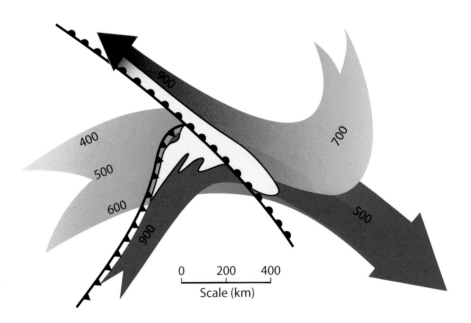

**Above** The three conveyor belts. The main rain-bearing flow is shown in dark blue.

# Blocking Highs

As described in Chapter 2 (page 69) the boundary of the northern polar front generally occurs as several (four or five) lobes around the Earth, with troughs of cold polar air extending towards the equator, and ridges of warm tropical air extending towards the pole. When the winds are strong, the lobes are small, but when the winds weaken the troughs and ridges become very pronounced in latitude, extending far to the south and north. Sometimes these waves become so extreme that they cease to move around the globe in the general westerly flow. The high-pressure areas, or anticyclones, become stationary, and may persist in a particular area for a long time. Anticyclones that become stationary for a long time are known as 'blocking highs', and tend to divert depressions from their normal progress towards the east. This forces the depressions north to higher latitudes, or south towards the equator. A blocking high over the eastern Atlantic, for example, sends some depressions south of their normal track over Western Europe, causing Spain and the Mediterranean to experience unusually changeable conditions. Other depressions might be diverted to the north, ending up over the Arctic.

The location of the blocking high plays a large part in the weather. This is particularly the case in Western Europe in winter, when a high settles over Scandinavia. (This high is usually an extension of the extremely cold Siberian High.) The easterly winds circulating around the southern side of the high drag intensely cold air over the region, bringing extremely cold weather. Britain experiences heavy snowfall because of the moisture that the air acquires as it passes over the North Sea. In summer, however, anticyclones do often produce areas of light winds, clear skies and long, hot sunshine. The light gradient winds mean that local winds (page 105) are often more pronounced. In autumnal anticyclones, in particular, extensive fog frequently forms. There may also be significant frosts at night, except during the height of summer.

# Anticyclonic Gloom

The descending air that is in the centre of anticyclones warms during its descent, suppressing the formation of clouds, and particularly of high-level clouds. If it completely suppresses all clouds and none whatsoever form during the day, the temperature will drop swiftly after sunset and we will have a clear, cold night. But the descending air also creates inversions at low levels that may block the growth of convective clouds. When that happens the clouds form a layer of heavy stratocumulus – an 'anticyclonic gloom' – that can hang around for days. Worse, the low-level inversion also tends to trap pollutants that lie close to the surface and near the centre of the anticyclone, where there is no wind to disperse them.

## Smog in London

In December 1952, pollution from vehicle exhausts and coal fires was trapped over the London Basin by an anticyclonic inversion. This led to London's Great Smog. The extreme conditions over several days resulted in many deaths – and later analysis suggested that an additional 4,000 people died prematurely and as many as another 12,000 suffered some form of respiratory illness. More recent research suggests the death toll was even higher. This tragic event eventually led to the Clean Air Act of 1956 and a later Act of 1968, which introduced areas where only smokeless fuels could be burnt. Though London's 'pea-soupers' are things of the past, dense 'clean' fogs still occur, and London suffers from a different form of year-round pollution from vehicle exhausts.

**Above** A policeman using a flare to direct traffic during a dense smog in 1952.

# Mist and Fog

Technically speaking, mist becomes fog when visibility drops below 1km (²/₃ mile). But in weather forecasting, visibility of less than 200m (565ft) usually qualifies as 'fog' and visibility of less than 50m (164ft) as 'dense fog'.

The calm conditions found in anticyclones are ideal for the formation of mist and fog.

There are two types of fog: radiation fog, which forms over land, and advection fog, or sea fog.

**Above** Mist at Stowe, Buckinghamshire.

## Radiation Fog

Radiation fog forms when clear skies cause the temperature at the surface to drop, through radiation to space. When the air at the surface reaches dewpoint, water vapour condenses (on the tiny particles that are present everywhere), creating what is essentially a ground-level cloud. This kind of fog may thicken overnight and is only dispersed the next day, when sunlight is able to penetrate the fog and warm the underlying ground. Radiation fog tends to form over low ground and often blankets valleys, leaving the higher ground in clear air.

The layer of fog, perhaps no more than a few tens of metres thick (typically 15 to 100m [49 to 328ft]), then often rises and forms a low-lying layer of stratus cloud. Valley fogs may often be seen flowing downstream and eventually dispersing. Sometimes a layer of cloud will have moved in during the night, so the next day's sunlight is unable to reach the ground and the radiation fog may persist through the following day. In industrial areas where pollutants mean there are high levels of particles in the air, the fog may also thicken early in the day.

Even a light wind at night will prevent radiation fog from forming. Above a speed of about 7.5km/h (around 4mph), the wind creates sufficient turbulence to stop its creation. Similarly, if the wind rises during the night, fog may disperse very quickly.

Hill fog will often form if a gentle wind carries humid air up the sides of hills, forcing the air to cool to its dewpoint. This kind of fog will often disperse to the leeward of the hills (the side sheltered from the wind), where the air descends and warms. Upland areas are particularly prone to be covered in 'Scotch mist', the name for a mixture of fog and drizzle, very common on Scottish hills but also in Devon and Cornwall, where it is known as 'mizzle'. If you are walking in upland areas you always need to be prepared for such conditions whenever there is a moist airstream that may condense into fog – which is essentially the same as a stratus cloud shrouding the highest areas.

**Above** Hill Fog in the Brecon Beacons National Park, South Wales.

## Fog that Moves

A very light wind will sometimes carry fog that has formed in one spot over to neighbouring, previously fog-free ground. This is particularly hazardous for drivers, when patches of dense fog suddenly appear, unexpectedly, ahead on the road. The horizontal transport of air (or fog) is known as advection. Advection may apply to radiation fog, although it is more commonly used to describe sea fog.

## Sea Fog

Sea fog, also known as advection fog, forms when humid air flows over cold water. The amount of fog depends on the amount of humidity in the air. At low humidity, fog may form only in patches, but when the air is very humid it creates a dense blanket over the sea. The wind then carries the fog over the neighbouring land. A similar sea mist or fog may also be produced when humid air from the sea (typically maritime tropical (mT), air) is forced to rise over the coastline. In either case, the wind may carry the fog a long way inland. In eastern Scotland and the north-east of England this cold sea fog is known as the 'haar'. It is particularly frequent in spring and early summer when the temperature of the water in the North Sea is low. (Sometimes the term 'haar' is used for just a the sea breeze itself, unaccompanied by fog or mist.)

Sea fog or mist may occur at any coastal location, whenever moist air overlies cool waters and there is a sea breeze to carry it inland (it doesn't need to be a gradient wind created by the difference between regions of high and low pressure). When the land is warmed by the Sun, a sea breeze (see page 105) will set in, and often carries sea fog or mist onto neighbouring coasts – spoiling conditions for people hoping for a bit of sunbathing on the beach. Where the immediate coastline is high, perhaps with cliffs – as found along the parts of the coast of Devon and Cornwall – the additional ascent over the higher ground causes the mist or fog to thicken and produce unpleasant, even potentially dangerous, conditions.

**Below** Fog at Rhossili on the Gower Peninsula.

# Dew and Frost

The quiet conditions found in anticyclones mean clear air, which allows objects on the ground to radiate heat away to space at night.

## Dew

Dew forms when water vapour in the air cools to its dewpoint and condenses as droplets (dew) on exposed surfaces. Much of the water vapour actually comes from the ground itself, which remains at a slightly higher temperature than objects that are higher up. Dewdrops are normally about 1mm in diameter. They may cause certain optical effects when illuminated by sunlight, such as dewbows and the effect known as 'heiligenschein' (see Chapter 5).

**Below** A dewbow formed on dew-covered grass.

## Frost

When, under clear skies, the air temperature hovers just above freezing, the ground itself may well be below 0°C (32°F). If there is a light breeze, most of the ground will be frost-free, but frost will form in more sheltered spots. If, on the other hand, the air temperature is also below freezing, the ground will be colder still by several degrees. Water vapour from the air then freezes directly onto exposed surfaces, without going through a liquid phase, creating a coating of hoar frost on trees and shrubs and icy patches on roads.

Sometimes dew will form on objects, and after this the temperature drops below freezing. The dew does not always freeze immediately, and the droplets may become supercooled (remaining liquid) below 0°C. Once any ice crystals have formed, these grow at the expense of the supercooled droplet, spreading in intricate patterns across the surface. This is what has happened when you see the familiar fern-like frost patterns on the windowpane.

Supercooled dewdrops will freeze spontaneously if the temperature drops to around -3 to -5°C (27–23°F). Normally such droplets are distinct, but if the dew has been very extensive they may combine to produce a thin sheet of ice on the surface. This film of ice should not be confused with true glaze (see page 56) that forms when rain falls onto surfaces that are well below freezing.

**Above** Hoar frost on the edges of fallen leaves.
**Below** A hawthorn tree covered in rime.

# Rime

When fog droplets are supercooled, they freeze immediately on contact with any solid surface. Although this may look like hoar frost, it is actually rime, which forms by a completely different method. This hard, white coating of ice builds up on the windward side of any objects with which the droplets collide. If there is a slight drift of wind, it may create long, icy 'feathers' that point into the wind. Or if the air is still, rime may form 'needles' of ice around the edges of leaves and other sharp objects. The crystals are usually much longer than those of hoar frost. Rime gives trees and other objects a 'fairytale' coating of ice, but may sometimes be so heavy that it causes damage to trees and shrubs, although far less damage than the glaze that produces occasional 'ice storms'.

**Below** Walkers at Wasdale, Cumbria.

# Water in the Atmosphere

Water exists in the atmosphere in numerous forms. It may be suspended in the air as mist, fog or clouds; deposited on the surface as dew, frost or rime; or exist as precipitation. Precipitation includes drizzle, rain, freezing rain, hail and snow (together with rarer forms, such as ice crystals and ice pellets). So you can see that the way that water, and especially water vapour, behaves in the atmosphere is extremely important to weather conditions.

Water is unique in that it may exist in three different forms (known as 'phases'): water vapour, liquid water and ice. Energy – that is, heat – is required to melt ice into liquid water, and also for liquid water to evaporate into water vapour. As mentioned previously (see page 53), when condensation or freezing occurs this heat, known as latent heat, is released into the parcel of air. Beware the common misunderstanding that air 'holds' water, as if it were a sponge. Water is certainly present throughout the atmosphere, but as water vapour rather than a liquid form. Water vapour is a gas, so it behaves like all other gases.

## Why Humid Air Rises

There is a specific physical law, Avogadro's Law, stating that the number of gas molecules in a given volume of air will be the same, so long as the pressure and temperature conditions are the same. Interestingly, this doesn't change for different kinds of gas molecules, nor for different sizes of molecules. In the case of the atmosphere, dry air consists of approximately 80 per cent nitrogen and 20 per cent oxygen (with some trace gases that can be ignored here). That means there are roughly four molecules of nitrogen to every molecule of oxygen. If (say) ten molecules of water vapour enter the air, eight of nitrogen and two of oxygen are forced to leave, to balance things up. However, the combined atomic weight of the gases expelled is greater than that of the water molecules that replace them. Humid air – containing more water vapour – is thus less dense than dry air. This is why a humid air stream will tend to rise above a dry one. An example is when hot, dry air from the Sahara flows north across the Mediterranean. When the hot, dry air encounters cold air, it initially hugs the surface rather than rising above the cooler air. But as it gains humidity from the sea, the hot air starts to rise.

# The Origins of Rain

Until the twentieth century and the growth of aviation, the actual processes by which rain formed were a complete mystery. Condensation into cloud droplets was understood, but not how they formed the raindrops that reach the ground. Cloud droplets are so tiny (about 20 micrometres across) that they remain suspended in the air, hardly drifting downwards under the influence of gravity. Even the smallest raindrops are about 2mm ($^8/100$in) in diameter – and that would take about one million cloud drops to create. (The largest known raindrops are about 6 mm [¼in] across – any larger and the drops break up into smaller droplets as they fall.)

Initial understanding came with the realisation that freezing (glaciation) is important. Once freezing has been initiated in the highest regions of a cloud, tiny seed ice crystals grow rapidly by capturing water vapour from the surrounding air. When the resulting, large ice crystals eventually fall into warmer layers, they melt and become raindrops. This overall process is known technically as the glaciation process or, less formally, as 'cold rain'. This is the process that produces most rain in temperate regions. So, if you notice that the tops of cumulonimbus clouds are becoming cumulonimbus capillatus

(see page 64) you can be sure that rain or hail is coming soon.

But this did not solve the problem. It was realised that heavy rain occurs in the tropics, where some clouds do not reach the freezing level. It was discovered that in very deep clouds there are two ways in which raindrops may form. In the most obvious, turbulence is so vigorous that it causes even the tiny cloud droplets to collide and combine. In the second, once initial droplets have formed and start to fall, they overtake and collide with smaller cloud droplets, coalesce with them and so become even larger. This coalescence process – sometimes called 'warm rain' – occurs not only in tropical clouds but sometimes in temperate regions in summer. In temperate regions, some 'warm clouds' that don't reach freezing levels – convective clouds, such as cumulus congestus and cumulonimbus – are still deep enough to harbour considerable turbulence. This promotes collisions between the falling particles, which become true raindrops and give rise to heavy showers.

**Above right**  Rain at Birling Gap, East Sussex.

# Precipitation

As mentioned earlier (pages 52–53) the three ways in which clouds arise are matched by three ways in which rainfall occurs: convective rain, orographic rain and frontal (cyclonic) rain. Convective clouds produce showers although, unfortunately, there are no obvious signs of when deep cumulus congestus are likely to produce rain. The best indication that those summertime clouds approaching you are likely to bring rain is if you can see that rain is already falling from them – look out for a ragged base and visible streaks of precipitation. The fallstreaks of rain (or snow) visible below the cloud, but evaporating before they reach the ground, are a kind of accessory cloud, known as virga – from the Latin for 'streak' or 'stripe'.

Most mid-latitude depressions that occur over Britain bring rain or snow, but they may bring other forms of precipitation, such as drizzle or hail. Another point to note is that rain also tends to bring higher winds at ground level. This is because the falling rain tends to carry the higher wind speeds found at altitude down towards the ground, increasing the gustiness and strength of the wind at the surface.

# Showers

Both cumulus congestus and cumulonimbus clouds produce relatively short-lived showers of rain. For cumulonimbus this happens in recognisable stages. The first stage, when the cloud is still cumulus congestus, lasts for about 20 minutes. Once the freezing level has been reached the transition is rapid. The period when the cloud is cumulonimbus calvus and cumulonimbus capillatus may be considered the mature stage, and also lasts around 20 minutes. Some large drops of rain may fall at the cumulus congestus stage, but most of the precipitation occurs in the mature period, initially as large raindrops when the cloud is still cumulonimbus calvus. Following this, there will be heavy rain, perhaps accompanied by hail, still in the mature stage. In the final (decaying) stage, rainfall gradually decreases in intensity, and the raindrops decrease in size. This can take anything between 30 minutes and about two hours. Wind may be a factor, because if there is strong wind shear (a great increase in wind strength with height), the top of the cloud may spread out downwind, and that may give some rain before the heaviest rainfall arrives, and also increase the overall duration of convection and rainfall.

Each cumulonimbus cell has a limited life of (depending on circumstances) between 1 and 3 hours, unless continuing convective activity causes several cells to combine into a single cloud mass. Different portions of the cloud cluster may be at differing stages of their development. If you are standing underneath the cumulonimbus you may experience one or more cells passing overhead, usually with the rainfall from one cell dying away to be replaced by rain from a younger cell, but with cloud cover remaining unbroken. When cells are accompanied by lightning (as discussed in Chapter 4), the approach of new cells can sometimes be perceived.

**Right** Rain on the South Downs, West Sussex.

# Updraughts and Downdraughts

When an ominous cumulonimbus shower cloud is approaching, carried by the prevailing wind, you may wonder why the surface wind sometimes appears to be flowing towards the cloud. In vigorous cumulonimbus clouds (in particular) the falling rain creates powerful cold downdraughts within the cloud itself. When these strike the ground, they tend to spread out ahead of the cloud. Sometimes they may be so strong as to cause a significant gust front ahead of the advancing cloud. The layer of cold air tends to undercut warm air that is being drawn into the cumulonimbus by the updraughts within it. It is this inward flow of air that forms the 'wind' towards the cloud, and which is often felt by observers, generally before the arrival of the cold gusts from the cloud itself. And this is why clouds sometimes appear to an observer to approach 'against the wind'. As the inflowing air is lifted away from the surface by the cold outflow it often forms a smooth-topped wedge of cloud – a shelf cloud – ahead of the cumulonimbus itself.

# Orographic Rainfall

When air is forced to rise as it encounters hills or mountains, it results in orographic clouds and rainfall – considerable amounts of rain or snow may result from even quite gentle winds. A common situation on the south coast of Britain in summer is when the sea breeze reaches the line of hills inland, such as the South Downs. Not only does cloud build up over the hills, but rain may fall on them. The additional lift provided by the high ground may sometimes lead to extensive flooding through orographic rainfall.

**Right** Heavy rain at Scafell, Cumbria.

# Flash Floods

**Above** Boscastle, Cornwall.

Flash floods may cause disastrous natural disasters. On 31 July 1976, an extremely vigorous cluster of cumulonimbus clouds built up over the Big Thompson Canyon in the Rocky Mountains of Colorado, USA, remaining stationary over the mountains and producing extremely high rainfall. The flood waters swept down the canyon, causing devastation, and 143 people were killed.

In 2004, Boscastle in Cornwall, a picturesque fishing village situated in a long inlet looked after by the National Trust, was hit by devastating floods. Sea breezes from both sides of the Cornish peninsula had combined and created extreme lifting over the high ground in the centre of the peninsula. This brought very heavy rainfall that was channelled into the valley of the River Valency and flowed down to Boscastle, which was wrecked by the sudden deluge. Fortunately, thanks to the exemplary work of the rescue services, no lives were lost, although in 1952, similar circumstances during the notorious Lynmouth flood took 34 lives in north Devon. In this case it is likely that heavy rain on a cold front was amplified by the orographic effect, with the rain falling onto already saturated ground.

When the wind carries humid air over hills or mountains, much of the rain or snow is deposited on the windward side of the upland area. To leeward, the air is depleted in moisture, leading to a 'rain shadow' and reduced rainfall on the far side of the hills. In addition, when air has lost a lot of moisture, it warms at the same, or greater, rate in its descent than the rate at which it cooled on its ascent. This may lead to what are known as 'föhn' conditions, where the air plunging down into leeward valleys becomes very warm and may, for example, cause the rapid dispersal of lying snow. It may also bring a fire risk to the area because of the desiccating effect of the extremely dry air.

# Frontal Rain

The amount and duration of rain or snow in a depression system depends on the stage of its development. Even at the earliest stage, when the depression is little more than a minor wave on the polar front, and before any distinct warm sector (see page 70) has developed, there will be rain ahead of the warm front that will extend round on the polar side to merge with rain on the advancing cold front.

In a fully developed warm-sector depression the rain ahead of the warm front may last between 6 and 12 hours. It will mainly fall from stratiform cloud (nimbostratus), with bands of heavier rain running roughly parallel to the surface front. Sometimes, convective activity ahead of the front may persist and active cells may be mixed in with the general cloud cover, giving rise to bursts of heavier rain.

Depending on the exact nature of any depression, there may be considerable rainfall in the warm sector itself, and some upland areas tend to receive more rain from the warm sector than from the warm front itself.

Sometimes there is considerable convective activity, with strong cells ahead of the actual cold front. This activity can be so strong that it produces a line of powerful cumulonimbus clouds a little ahead of the frontal zone – known as a squall line. In general, cold fronts are marked by instability, active convection and the corresponding clouds, which are quite unlike the relatively quiet layer clouds formed by the gradual ascent of warm air at the warm front. The rainfall is of shorter duration, although it is often very heavy and may be accompanied by thunderstorms.

Rainfall at occluded fronts is generally similar to that at warm fronts, and although the band of rain is usually narrower, it may still be heavy. There is little convective activity. Occluded fronts may still bring prolonged periods of rain, particularly if the path of the depression means that the front is overhead for a long time.

# Drizzle

It is possible for cloud droplets to coalesce even in shallow clouds, and without there being vigorous convection and turbulence. This can happen when warm, moist, stable air (such as that in the warm sector of depressions or, indeed, in any warm air stream) is forced to rise over high ground. As it cools it is subject to gentle turbulence, and this is when small droplets (no more than about 1mm [$^4$/100 in] across) may sometimes form. These create damp drizzle or, if the air is cold enough, a light fall of snow or even tiny ice crystals. Drizzle commonly

occurs beneath a layer of thick stratus or stratocumulus cloud. In combination with hill fog, and because of its prevalence in the Scottish Highlands, it is sometimes referred to as 'Scotch mist' – a slightly ironic term, given that it is continuous drizzle rather than mist.

The tiny ice crystals (again less than 1mm across) that occasionally fall from the same kinds of cloud that produce drizzle are known as snow or ice grains. In North America these are sometimes called sleet, not to be confused with the British use of the term to mean a mixture of freezing rain and snow.

**Right  Snow at Kirkstone Pass, Cumbria.**

# Ice Crystals, Snow and Soft Hail

The exact form of the ice crystals that arise in the tops of cold clouds depends on the temperature: the colder the temperature, the smaller the ice crystals. At temperatures just under freezing, the crystals form with six arms or spikes – the classic 'snowflake' shape we know from Christmas cards and school drawings. The crystals are called dendrites, from the Greek for 'tree'. In reality, actual snowflakes that reach the ground consist of multiple dendrites, entangled and trapped together. There are other types of ice crystals that form at lower temperatures, including thin flat plates, needles, hollow columns, and flat, branched plates.

When falling snow crystals of any type collide with supercooled water droplets, the two freeze together, in the process known as riming, and may eventually turn into small, round pellets of hail, known as graupel or 'soft hail', about 2–5mm (up to $1/5$ in) across. These pellets most commonly form in cumulonimbus clouds in cold weather. This soft hail sometimes reaches the ground, but the pellets are much smaller and softer than true hail, which forms by a different mechanism that will be described later in the context of severe weather and thunderstorms (see pages 119–120). If unrimed crystals fall into warm air, they melt slightly and are held together to form the much larger, true snowflakes that fall from the clouds.

# Local Winds

As with the names of winds in general (such as the easterlies or westerlies), there are local winds that are named after their place of origin. So the sea breeze begins at sea and the land breeze over land; lake breezes start over lakes; valley winds flow up valleys and mountain winds blow down towards the lowlands. But if the regional gradient wind is strong, any of these local winds may fail to develop.

## The Sea Breeze

Particularly in spring and early summer, when the air-stream is humid and the sea-surface temperature is low, sea mist or fog (see page 89) may be present, and may be carried far inland by the sea breeze. A fine, clear morning may turn into a damp, misty afternoon, spoiling your plans for a sunny day out. But on a high summer day at the beach, the day may start calm and warm, but by afternoon a wind has set in from the sea.

This wind, often bringing welcome relief from high temperatures, is a sea breeze. It is caused by the land absorbing heat more rapidly than the sea, which produces convection and rising air above the land. This in turn creates a gentle pressure gradient, drawing in cool air from the sea. This cooler air is normally confined to a layer that is perhaps no more than 3,000ft (915m) deep, above which there is a corresponding flow of air moving out towards the sea, completing the circulation. Sometimes the seaward flow at height produces a thin sheet of cloud, which may be seen gradually extending out to sea.

You might expect that a sea breeze would be at right angles to the local coastline, but there may be considerable deviation from this direction. This deviation is dependent on a variety of factors, such as the direction of any overall gradient wind, the local topography (with the breeze tending to favour lower-lying land), and the location of the area of land where heating is at a maximum.

There is often a definite sea-breeze front, marked by a distinct change in

**Left**  A breezy sea at Kimmeridge Bay, Dorset.

air temperature. If the countryside is fairly flat, the front may sweep as far as 40– 50km (25–31 miles) inland. The front is often marked by a line of convective cloud (usually cumulus, although occasionally cumulonimbus arises) and, as mentioned previously (see page 53) may give increased cloud or even rainfall if it encounters a line of hills.

## The Land Breeze

Anyone living near the sea will almost certainly know the morning breeze off the land that dies away and is replaced by the sea breeze later in the day. During the night, conditions are reversed from those in the day, and the land cools more quickly than the sea. The cool air moves out to sea in the land breeze (with a corresponding landward flow at height). The outward breeze may extend a considerable distance from the shore and be particularly noticeable to sailors.

As with the sea breeze, there may be a distinct land-breeze front that is marked by convective cloud, which is often readily visible on satellite images.

## Lake Breeze

Large bodies of water may also produce breezes similar to sea breezes, although here the situation is complicated by the nature of the surrounding land and the orientation of the lake, especially if it lies in a valley. If the water extends east–west and there are hills on the northern side, the hillsides will warm rapidly in the sunshine, creating a breeze on one side of the lake. If, on the other hand, the lake lies in a valley that runs north–south, or the hillsides are in shadow, any lake breeze may be very weak or absent.

# Valley Winds

Heating of hillsides during the day tends to create a current of air that flows up towards the head of a valley and also up the slopes towards the ridges on either side. These valley winds tend to arise early in the day as heating begins. They reach their maximum strength (perhaps 20km/h [12½mph]) shortly after midday, and then die down at night, when they may be replaced by a mountain wind. The maximum speed is reached over sun-warmed slopes, but there may be little (or no) wind over areas in shadow.

You can sometimes see visible signs of the wind in cloud. If a thin valley fog formed overnight, there may be sufficient heat reaching the ground early in the day for the fog to lift into a low layer of thin stratus cloud. This often then breaks up into small patches, which may sometimes be carried up the sides of the valley.

# Mountain Winds

A mountain wind is the counterpart to a valley wind. It builds up when the heating by sunlight starts to decline, late in the day, and cool air begins to slide down the slopes towards the mouth of the valley. Such mountain winds tend to be weaker than the corresponding valley winds, reaching maximum speeds of some 12km/h ($3^2/_3$ kph), but the speed may increase if the sides of the valley close in to form a narrow gap or ravine. Mountain winds tend to die down shortly after sunrise, although early morning remnants of overnight fog are often seen drifting down the valley, carried by the end of the mountain wind.

Both valley and mountain winds may be modified or even prevented by the overall gradient wind. The strongest winds arise under high-pressure, anticyclonic conditions, with clear skies providing strong heating during the day, and allowing rapid cooling at night.

**Below** Windy Mam Tor in Derbyshire.

# Katabatic Winds

The term 'katabatic winds' may seem odd, but it aptly describes cold winds that plunge down mountainsides. You may remember, in relation to subdued cold fronts, that the meaning of the Greek word 'kata' is 'down' (see page 81). Winds like this, that remain cold – unlike the downslope (föhn) winds that become very warm – are sometime called 'fall winds'.

When cold air builds up over a highland region, especially when that is covered in snow or ice, it becomes extremely dense. It cascades down any slopes, and may be exceptionally violent. The most extreme examples are the frigid katabatic winds that spill down from the icecaps of Greenland and Antarctica, but Europe has a major example in the Bora, that rages down from the Dinaric Alps over the Dalmatian coast, and then out over the Adriatic Sea. (This wind is so violent that the roofs of buildings in areas under its path are especially constructed to withstand the blast, which often creates damage and overturns vehicles.)

The only named wind in Britain, the Helm Wind, is of this sort. This is a ferocious, bitter wind that cascades with a roaring sound down into the Eden Valley in Cumbria from the escarpment west of Cross Fell, the highest summit of the northern Pennines. It occurs only when there is high pressure over the northern North Sea, producing a north-easterly wind at Cross Fell. Acorn Bank, the National Trust property in the Eden Valley, is sited, like many other buildings in the area, so that it turns its back on the violent, shrieking Helm Wind. The Helm Wind is accompanied by a bank of cloud (the Helm) that covers Cross Fell and lies along the top of the escarpment. A further bank of cloud (the Helm Bar) generally lies parallel to the Helm and several kilometres downwind. Below the Helm Bar the wind may be absent or even blow towards the mountains. The Helm Bar is a form of wave cloud (see page 65), where the normal conditions promote the formation of a single, major wave downstream of the peak, although multiple bars have also been noted.

**Left** Acorn Bank in Cumbria, with the high fells behind, the source of the penetrating Helm Wind.

# Föhn Winds

The Helm Wind in Cumbria, as with any downslope wind, becomes warmer as it descends and the air is compressed. More extreme examples are the föhn (or foehn) winds that affect many regions of the world, including the regions north of and below the Alps, where the term originated as a generic term for these hot, southerly winds.

Air ascending the windward slopes of the mountains not only reaches the dewpoint but also releases large quantities of water in the form or rain or snow on the windward slopes of the mountains. When the air passes over the summits and descends to leeward, it warms at a much greater rate than the rate at which it had cooled on its ascent. Having by now lost so much moisture, it is very dry. Its desiccating effect on standing crops and wooden buildings is so great that it may even create a fire hazard. The arrival of the föhn may be accompanied by an extreme jump in temperature.

## Chinook Wind

Föhn winds are also noted for causing the rapid thaw of lying snow. The Chinook wind in the Pacific Northwest and inland United States is also known as the 'Snow Eater', for this very reason. As it descends the eastern side of the Rocky Mountains it may cause rapid temperature changes. On 23 January 1943, a Chinook at Spearfish, South Dakota, brought a huge temperature rise of 27°C (81°F)(from -20°C to 7°C [-4°F to 45°F]) in just two minutes.

# Local Effects

Apart from the specific, named winds just described, there are a number of other effects that may occur, often influenced by nature of the landscape.

## Sailing Winds

Sailors are particularly conscious of the effect of steep mountainsides or cliffs in channelling the wind and thereby increasing its strength, particularly in certain parts of the world. The rocky cliffs lining the strait between Corsica and Sardinia have been known to increase a relatively light breeze into a gale-force wind. A similar effect occurs in the Straits of Gibraltar. Here, a westerly wind may double its velocity, but the effect is even worse with a strong easterly wind. Not only is there an increase in the wind speed, but the wind also acts on the surface water, which exhibits a permanent current (an inflow) to the Mediterranean from the Atlantic. The effect of 'wind against water' produces choppy waves and unpleasant sailing conditions.

## Channelling the Wind

There is always a tendency for the wind to follow the course of a valley, in particular if the valley is deep. A similar 'wind channelling' effect occurs in towns and cities, where the effect of buildings may be to channel the air in a particular direction and produce strong gusts. The wind may also be channelled over the cols that lie at the heads of valleys. (A col is the lowest point of a ridge between two peaks.) Such gaps in a range of hills or mountains are often known as 'wind gaps', because the effect is so marked. This is particularly the case when the line of hills or mountains lies roughly at right angles to the prevailing wind.

Wind speeds may be increased considerably if a valley narrows into a gorge – there are a few such 'ravine winds' that are so frequent that they have been given specific names. The Kosava, for instance, which occurs where the Danube cuts through the Carpathian Mountains. Narrow ravines that funnel the wind can give rise to various minor whirlwinds ('whirls'), described in Chapter 4.

# Wind and Coastlines

As already seen, the strength and direction of the wind changes when it crosses a coastline and moves from water to land, or vice versa. Even if the wind is parallel to the coast, however, its speed will be greater over the water than over the land because of the lessened friction. When there is a headland, pointing more or less downwind, there is a tendency for the wind to become stronger just offshore, and for the wind to curl round the headland, creating an eddy behind it. If the coast is particularly steep, a distinct low-pressure area may develop behind the headland, away from the open sea.

If the wind is onshore and the coast is steep (perhaps with cliffs), and the air is unstable, there may be sufficient additional uplift given to the air to produce cumuliform cloud along the coast. If the airstream is already showery, the additional lift may be more than enough to turn the showers into fully fledged thunderstorms. The effects of a steep coastline are not confined to its immediate vicinity. The changes may occur 5–10km (roughly 3–6 miles) out to sea, and prevail as much as 50km (31 miles) inland. In summer, a strong onshore wind acting with a steep coastline may result in thundery conditions at night, when convection might normally be expected to die away.

A steep coastline may create other phenomena. If the wind is off the land, there will be a relatively calm zone stretching out to sea with little, or no, wind. The zone may extend as far as about ten times the height of the cliffs. The combination of steep cliffs and an offshore wind can result in the formation of a large eddy (known as a 'pillow eddy'). At its base the surface wind is gusty and blows towards the coast. Or, if the wind is in the opposite direction, a different eddy (a 'bolster eddy') may form against the cliff, again creating a surface wind that is in the opposite direction to the main gradient wind. Similar eddies can form inland, when the wind encounters steep scarps.

**Left** St Catherine's Oratory, known affectionately as The Pepperpot, on St Catherine's Down on the Isle of Wight, is the only surviving medieval lighthouse in the UK.

**Pages 114–115** Forked lightning over the Clifton Suspension Bridge, Bristol.

# 4. Severe and Unusual Weather

Luckily, Britain does not experience the extreme weather events that occur elsewhere in the world. Severe weather, such as major thunderstorms and exceptionally strong winds, does still take place. Even quieter weather may occasionally be accompanied by unusual, or even rare, phenomena that are extremely striking to observe.

# Thunderstorms

Thunderstorms develop from large, active cumulonimbus clouds. Even today, there is no generally accepted theory of how the positive and negative electrical charges become separated into the various regions within these clouds. We know that the process occurs only when glaciation (freezing) happens at temperatures of approximately -20°C (-4°F), and that both water droplets and ice crystals must be present. One theory is that the charge separation (the building up of space between particles that have opposite charges) occurs when water droplets freeze. During the process, the ice crystals fragment. The lighter fragments carry a positive charge and these are swept up to the top of the cloud, while the heavier, negatively charged particles fall to the base of the cloud. An alternative theory is that the separation of charges occurs when ice crystals collide, when the smaller particles receive a positive charge.

**Above** Stormy skies over West Wycombe Park, Buckinghamshire.

# Lightning

However the separation happens, the result is that the top of the cloud acquires a positive charge, and the base a negative one. The negative charge at the base of the cloud causes an opposite, positive charge to develop on the ground beneath it. This positive charge follows the cloud as it is carried across the landscape. Eventually, either the charge becomes so great or the distance separating the cloud and the surface charges becomes so small (for instance, such as over a tall tree or high building) that the electrical resistance breaks down, giving a cloud-to-ground lightning discharge.

Discharges may also occur between different regions of the same cloud (intracloud lightning), or between two clouds (intercloud lightning). Again, the exact trigger to any discharges remains unknown; one suggestion is that they are initiated by cosmic rays from space. Many different electrical phenomena, linked to lightning, have been discovered in recent years, and these have been given a whole range of both prosaic and exotic names. These include 'blue jets', 'blue starters', 'sprites', 'elves' and 'trolls'. most of which are too brief to be seen by human eyes. Other phenomena associated with lightning include bursts of X-rays and high-energy gamma rays, and even showers of positrons (the antiparticle counterpart to electrons). The various mechanisms behind the production of all these phenomena remain obscure.

## Saint Elmo's Fire

The phenomenon known as St Elmo's fire happens when electrical charge can cause a continuous glowing discharge from sharp objects, such as trees, church spires, flagpoles and (at sea) from masts and the ends of spars. It may even occur at the ends of fingers or hair. The 'fire' – often blue or violet in colour, and sometimes appearing very much like flames – is actually a kind of plasma caused by the electrical field around the object. St Elmo was another name for St Erasmus. He is the patron saint of sailors, who often took the sight of St Elmo's Fire on a ship as a good omen. While St Elmo's fire itself is not lightning, if your hair starts to stand on end, a lightning strike is extremely likely, and you should drop immediately to a crouch – as described in the section on thunderstorm safety.

# The Sound of Thunder

Thunder is the sound caused by lightning. A normal lightning discharge is extremely hot – far hotter than the surface of the Sun – and the channel of heated air expands and then contracts at supersonic velocities, creating the noise of thunder. Nearby lightning will be heard as a sharp crack, while the sound from more distant lightning discharges will be heard as a booming roll of thunder, as sound from different parts of the channel arrives at different times.

## How Close?

Measuring the time between seeing lightning and hearing the thunder is a fairly well-known trick to telling how close the storm is. This is because light from the flash travels much faster towards us than the sound. Counting the seconds between seeing the flash and hearing the thunder gives a rough estimate of our distance from the discharge. It works out at about three seconds per kilometre (or five seconds per mile). If you see lightning but don't hear any thunder, the thunderstorm cell is almost certainly some 20—30km (12½–18⅔ miles) distant.

You can estimate whether the lightning will pass overhead by noting the bearing of the discharges. If this remains the same from one flash to the next, then the cell is heading directly towards you. However, the lifetime of any one active lightning cell is 20–30 minutes, so the activity may cease before it reaches you.

Colloquially, we tend to use two terms to refer to lightning: 'fork lightning' (when the actual discharge is visible) and 'sheet lightning' (when the discharge is invisible). In reality there is no difference in the discharges. It is just that for 'sheet lightning' the discharge occurs within a cloud or between two clouds, and the channel is hidden by cloud. It is also common to hear people refer to 'heat lightning', as if it is a kind that only happens in the high temperatures of summer. But 'heat lightning' is simply distant lightning that is so far away that there is no thunder to be heard at the point from where you are observing it.

There is a rare, and puzzling, form of lighting known as 'ball lightning'. This takes the form of a glowing spherical region of air, which may drift around, either outdoors or even within buildings or aircraft. Its existence was doubted for many years, and there is no satisfactory explanation for its formation, so any reports or photographs are of considerable scientific interest.

# Supercells

There is an even more extreme form of thunderstorm known as a supercell. Instead of the convective activity being organised in individual cells, a single, giant system arises. This consists of an enormous, rotating column of rising air, known as a mesocyclone. The top of the cloud may reach anywhere between 8 and 15km (approximately 26,000 to 50,000ft) in height. The system contains a complex set of updraughts and downdraughts, and cool air enters the system at middle levels.

As previously mentioned, in normal cumulonimbus clouds the downdraughts tend to suppress the rising air currents, which leads to the limited lifetime of any individual cell. In a supercell the rotation of the mesocyclone isolates the downdraughts and prevents them from quenching the updraught. This means that systems may become very long-lived and may travel a long distance across country. The circulation of air within a supercell is such that the strongest updraught produces a giant 'vault' within the system. This is ideal for the formation of very large hailstones. They may be carried upwards several times by the strong updraught, passing through levels where water droplets exist, getting bigger each time as they acquire a new layer of ice. (If you cut

**Below** Supercell clouds at Southport, Merseyside.

open a very large hailstone, layers of clear and opaque ice are clearly visible.) The hailstones eventually become so heavy that they fall out of the cloud, by now so huge that they may devastate standing crops and cause considerable damage to objects such as cars. Worse still, if the hailstones fall into one of the downdraughts, they may be projected downwards with considerable force and cause even greater damage on the ground.

Supercells are the most powerful storms and frequently bring torrential rain, damaging hail and frequent lightning strokes. They are, occasionally, the source of highly destructive true tornadoes (see page 128). They are relatively uncommon in Britain, although they sometimes originate over France or Spain, cross the English Channel, and may, on occasion, penetrate far inland. They are much more common over the Great Plains of the United States, especially when there is a deep pool of unstable air and vertical wind shear (the wind direction changes drastically with height), often persisting for many hours and travelling long distances across country. But it's not over yet; when supercells or multicell storms finally decay they sometimes send a surge of cold air across the surface. The cold air undercuts warm humid air and may in turn initiate a squall line with a long active core of convection. This linear shape persists as the squall line advances, although small intense cells tend to propagate backwards from the main active region.

## Named Storms

In September 2015, the UK Met Office and Met Éireann (the Irish meteorological service), announced that, from then on, names would be given to potentially devastating storms to help with recognition by the general public, following a similar process already used in North America. The criterion for giving a specific storm a name was a belief that there it would have a significant impact on conditions over the British Isles. In particular, high winds should be expected. In fact, the very first winter that this scheme was implemented brought some exceptionally fierce storms, with high winds and extremely heavy rain, and resulted in destructive floods. On 4–6 December 2015, Storm Desmond brought exceptional rainfall to Cumbria, during which Honister Pass gained the newest British rainfall record of 341.1mm (almost 13½in) of rain in 24 hours.

# Hurricanes

Any discussion of severe weather should perhaps include a brief mention of hurricanes. Regrettably, the media often describe severe storms in Britain as 'hurricanes'. This was especially true for the great storm of October 1987, when the media almost universally wrongly used the word 'hurricane' and even books were published using the word.

A true hurricane can never occur in Britain, because one condition for a hurricane's formation is a sea surface temperature of at least 27°C (81°F).

Anyone who has paddled or swum in the northern North Atlantic Ocean will know that it is always colder than that. Such temperatures are found only in the tropics, which is where the family of storms known formally as tropical cyclones originate. The term 'hurricane' is used for storms in the Atlantic, Caribbean and Eastern Pacific. The same kinds of storms are called typhoons in the Western Pacific Ocean

**Above** Devastated trees after the 1987 storm at Toys Hill, Kent.

and cyclónes in the Indian Ocean. In general, northern-hemisphere hurricanes initially track westwards, before undergoing what's known as a 'recurvature', when they veer north and northeastward. By this stage they are usually beginning to decay. Occasionally, however, they may combine with an existing frontal wave or a fully fledged depression (otherwise known as an extratropical cyclone), and will cause it to deepen dramatically. If the central pressure of such a depression drops by at least 1hPa per hour for 24 hours, the system becomes known as a 'bomb'.

## How powerful?

Hurricanes are extremely powerful storms. A single hurricane may release energy equivalent to half the total electrical energy production worldwide.

## Hurricane-force Winds

Britain and British waters may experience 'hurricane' force winds, but these are not hurricanes. On the Beaufort Scale, these are winds that have a speed of 32.7m/s (107ft/s) (approximately 118km/h [73$\frac{1}{3}$mph]) or more. Such wind speeds are rare. One instance occurred on 10 January 1993, in what became known as the Braer Storm (after the oil tanker broken up by the storm off Shetland). This extratropical cyclone was one that developed as a bomb. It brought violent winds to Shetland and northern Scotland, together with the record estimated low pressure of 914hPa in the central northern Atlantic.

## The Beaufort Scale

The Beaufort wind force scale, which goes from 0 to 12, is an empirical measure of wind force, using observations of how the wind at certain speeds affects the environment. For instance, at wind force 0 or 'Calm', smoke will rise vertically, while at wind force 8, or 'Gale', twigs will break off trees and impede progress. Wind force 12 is 'Hurricane' and will bring devastation.

**Above right** Lightning viewed near Bracknell, Berkshire.

## ⚡ Thunderstorm Safety

It is sensible to take precautions when there is the likelihood of thunderstorms and lightning. Hot, humid weather with active convection are signs of great instability, indicating that thunderstorms may develop. Certain middle-level clouds, in particular altocumulus floccus and altocumulus castellanus, are also an indication of future major instability and possible storms ahead. Cumulus clouds that reach the level of the altocumulus will gain a great deal of energy, and rapidly become giant cumulonimbus clouds that can develop into thunderstorms.

If you are out tracking weather and catch sight of lightning or hear the sound of distant thunder, take this as an indication to begin thinking about your safety. Electronic lightning detectors are available and will indicate when lightning is nearby, but it is sensible to try to track any active cells yourself, to see whether they are likely to come close to your position and whether it is necessary to take shelter.

Stay low. Lightning generally (but not always) strikes the highest object in the locality and also tends to be attracted to anything that is a good conductor of electricity or to sharp objects. So it is sensible to avoid being the highest object in the vicinity. Avoid or leave high locations, such as the tops of hills or mountains, and mountain ridges. Be aware that anywhere on or in the water is also dangerous, because one is likely – even if swimming – to be the highest object in the immediate area. (Sailors need to ensure that their craft are adequately protected against lightning.) Wire fences and wet ditches that would conduct electricity should be avoided. And, as everyone probably knows by now, it's unwise to shelter under isolated trees, to be carrying an umbrella or swinging a golf club. Come back later to get that hole in one.

If you are caught outside, sheltering in thick timber, and preferably under shorter trees, is reasonably safe. If in open country, crouching down – do not lie down – in the lowest possible spot, such as a dry (not wet) ditch, will help to protect you.

## Why Crouch for Safety?

If you are outside in a thunderstorm, crouching down can help you to avoid being struck by lightning. You need to be low, and to cover as small an area as possible. Crouch on the balls of your feet to minimise contact with the ground, wrap your arms around your knees and tuck your head down to your knees. This position minimises the risk that, if you are unlucky enough to be struck, the current will pass through the centre of your body and affect your heart. Instead, a lightning strike will run across the surface of your back on its way to ground – especially if your clothing is wet. It may do no more than leave a fern-like pattern on your skin that will fade within a few hours.

If you are inside a big building, you are reasonably safe, especially if the building has a steel frame or fitted with a proper lightning conductor. Similarly, a vehicle that has a metal body provides a high degree of safety. Inside a house, ideally you should disconnect all electrical appliances, such as television sets and computers; stay away from doors, windows and chimneys; avoid contact with plumbing and certainly do not attempt to take a bath or shower. Finally, do not use a landline telephone, and if possible disconnect it.

## A Bolt from the Blue

The expression 'a bolt from the blue' has a basis in reality. Sometimes a discharge will strike many kilometres away from the parent cloud, extending out and then turning downwards from apparently clear sky. The exact causes are unknown, but this phenomenon is yet another reason for being cautious when lightning is about.

# Unusual Events

No discussion of the weather is complete without mention of some events that may be rare but are nonetheless sometimes very dramatic. These include a whole family of whirlwinds, unusual clouds and even (although not, strictly speaking, weather-related), aurorae.

## Whirls

There are a number of phenomena that may be classified as 'whirls'. Some of these are relatively weak, but others (particularly true of tornadoes) may be exceptionally violent and bring considerable destruction.

- Devils
- Funnel clouds
- Tornadoes
- Waterspouts and landspouts
- Gustnadoes

# Devils

The simplest and most common devils are whirls created by the wind that funnels between buildings in a town or city. You will have seen how they can lift litter, leaves and pieces of paper into a whirling column of rubbish. In the country, similar whirls can arise when the wind funnels between the sides of a valley or narrow gorge. These can happen at any time of year and are also the cause of water devils, if the whirl crosses a lake, or snow devils, if it crosses snow-covered ground.

When the weather is hot, particularly in summer, you sometimes see similar whirls that are caused by intense heating of the ground, lifting dust and loose debris. These are often called dust devils, although they have various local names in different countries. The hot surface creates strong convection and a rotating column of air as the warm air at the surface pushes into the cold air above it. The strength of the whirl can vary according to the nature of the surface, and they are more likely to be visible over flat, hot surfaces when the surface is particularly dry with loose soil or other light materials – hay and straw devils are also relatively common. Dust devils may travel a considerable distance across the countryside, pushed forward by the momentum of the spin and gathering or sustaining energy as they move over a new hot surface. Under calm conditions they may remain stationary and, generally, tend to dissipate quickly. Dust devils rarely cause any major damage, though they pose a hazard to skydivers, as they can cause a parachute to collapse without warning. In some conditions, such as arid desert areas, they can become quite big and impressive. Sometimes a small cumuliform cloud may form at the top of the devil.

**Above** A dust devil.

# Funnel Clouds

Funnel clouds are rotating funnels of condensation, usually a dark grey colour, that descend from a cumulus congestus or cumulonimbus cloud towards the ground, but do not reach it. They look very like a tornado, but are not – although in certain cases they may become one. Usually they are short-lived and weaken soon after formation.

Funnel clouds are caused by strong convection (air circulation) inside cumuliform clouds and arise from condensation caused by the reduced pressure in the mesocyclone, the rotating column of air, or vortex, within the cloud. They are known technically as 'tuba' clouds. They are not particularly rare, and are often seen hanging below a layer of cloud, such as stratocumulus, that is hiding the deeper, active convective cloud from sight.

**Above** A funnel cloud in Cumbria.

# Is it a Tornado?

A funnel cloud that touches down on the surface becomes a tornado at that point. If it reaches the sea surface rather than land it is called a waterspout. There are some 30–35 tornadoes a year in the UK, but most are small and do little or no damage. The longest track on the ground for a British tornado is 107km (66½ miles) on 21 May 1950, which travelled from Little London in Buckinghamshire to Coveney in Cambridgeshire. It then lifted and became a funnel cloud, travelling another 53km (32 miles) to Shipham in Norfolk). Reports from along the track suggest that this was a single phenomenon, not a sequence of funnels, tracking beneath a particular storm.

# Waterspouts and Landspouts

Both waterspouts and landspouts arise from the same mechanism as funnel clouds. They are a relative of the tornado since they are, in effect, funnel clouds that touch down.

There are two types of waterspout – non-tornadic and tornadic, but it is the non-tornadic form that will be described here. These kinds of waterspouts are sometimes called 'fair-weather' waterspouts, and are by far the most common. Waterspouts form over the sea, or other bodies of water. Unlike tornadic waterspouts, which are usually a tornado over land that has moved to the sea, fair-weather waterspouts are very short-lived. They are associated with flat-bottomed cumulus towers, and are most often found in tropical and subtropical regions.

As it starts to form over water , a waterspout appears as a 'dark spot' on the surface, where the strong downdraught from the cumulus cloud touches the water. This is normally followed by the dark funnel cloud itself. This dark grey downdraught is surrounded by a pale, rotating updraught, which is often invisible. The inflow at the surface may sometimes be visible as a spiral on the surface of the water. Once the funnel has touched down, it usually throws up a cylinder of spray, known as the 'bush'.

Waterspouts are particularly likely to form when the sea surface temperature is greater than that of the air – for

instance, when cold air arrives behind a cold front, creating unstable conditions. They tend to appear as a group, with several visible (along with funnel clouds) at the same time. The average lifetime of an individual waterspout is about 15 minutes and typical diameters are 15–50m (29–164ft). Possibly the tallest waterspout ever recorded was over 1,500m (4,920ft), off New South Wales, Australia, but in British waters a waterspout reached 1,000m (3,280ft) (over the Solent). Both the diameters and heights of waterspouts are very variable.

**Above** A waterspout seen from Brighton beach, East Sussex.

## Are Waterspouts Destructive?

Waterspouts are not particularly destructive, unless you sail right into them – which some foolhardy sailors have been known to do. In the UK, it's thought they may have contributed to the Tay Bridge disaster of 28 December 1879 (when the bridge over the Firth of Tay in Angus, Scotland collapsed while carrying a train during a storm) because two or three were seen near the bridge.

# Gustnadoes

The delightfully named gustnadoes are often called 'twisters', and are a form of whirl that may cause some damage. These invisible whirls are caused, like waterspouts and landspouts, by exceptionally strong convection.

They often accompany gust fronts and very strong winds, but there is no condensation funnel cloud. Gustnadoes are commonly associated with the high winds in tropical cyclones (cyclones, hurricanes, and typhoons).

# Unusual Clouds

Clouds are rare in the stratosphere and the still higher mesosphere but there are some unusual clouds that show up in each region, called nacreous clouds and noctilucent clouds, respectively.

## Nacreous Clouds

Nacreous clouds, also called 'mother-of-pearl clouds' (nacre is another name for mother-of-pearl), are known technically as 'polar stratospheric clouds' (PSC).

These rare clouds crop up high above the height of tropospheric clouds, at about 15–30km ($9^{1}/_{3}$–$18^{2}/_{3}$ miles), and are visible only when the ground is in darkness, after sunset or before sunrise. After sunset, they glow brightly with strong iridescence (see page 151), in the evening moving from pastel shades that gradually shift to more fiery orange and red tints as time passes. Before sunrise, the colours change the other way, from fiery to pastel colours.

**Right** Nacreous clouds.

These spectacular clouds are a form of wave cloud (see page 65), and often show a typical wave-like structure. Because the atmosphere is so dry at the high altitudes at which they occur, temperatures below -83°C are required for them to form. At this temperature water vapour freezes into nitric acid trihydrate particles. Suitable conditions are more common over the Antarctic than over the Arctic region, but the clouds may be seen from the British Isles.

Despite their beautiful appearance, the clouds are not necessarily a good sign. Unfortunately the particles on the nacreous cloud surface provide ideal conditions for the chemical reactions that create ozone holes (see page 26). The CFC compounds derived from pollutants from Earth interact with the nacreous clouds, and this creates chlorine compounds, which destroy the naturally occurring ozone in the atmosphere. The clouds are more common in the Antarctic, which is why the southern ozone hole is more frequent and stronger than the one in the north, which tends to be infrequent. In both cases, the clouds generally first appear in spring, as sunlight returns to that region of the atmosphere after the long polar winter, triggering the formation of the clouds.

## The Scream in the Sky

There have been recent suggestions that nacreous clouds may have inspired the fiery clouds in the background of Edvard Munch's painting, *The Scream*. Another theory is that the dust and particles released into the atmosphere following the eruption of Krakatoa, in August 1883, caused strange red twilights that were observable in Oslo, Munch's home town, for some time. Perhaps we'll never know – in his journal, Munch wrote of being inspired by seeing a blood red sky that was like 'a scream passing through nature'.

**Right** *The Scream* by Edvard Munch, 1893.

## Noctilucent Clouds

The highest clouds in the Earth's atmosphere are noctilucent ('night shining' in Latin) clouds (NLC). They occur at the level of the top of the mesosphere, the mesopause (see page 25), at an altitude of about 86km (53½ miles). These clouds are seen in the middle of the night for a few weeks on either side of the summer solstice, when looking toward the poles. In Britain they are only visible when the observer is in darkness and the Sun is below the northern horizon. Because they are normally visible only to observers at high latitudes (above 45°N or S), they are sometimes referred to as polar mesospheric clouds (PMC). There are unconfirmed suggestions that they may also exist at lower latitudes and be fleetingly visible at sunset and sunrise.

Noctilucent clouds have a distinctive 'electric' (blue-white) tint that may become slightly yellower in shade, later on in the night. They look a little

**Above** Noctilucent clouds over Northumberland.

like high-altitude cirrus, but not as wispy – they are more of a very thin, undulating layer – and are made up of tiny ice particles about 0.1μm (1μm = one millionth of a metre) in diameter. People tend to speak of there being different 'types' of noctilucent cloud, due to their varying appearance, but in reality the differences are due to the differing length of the observer's line of sight through the undulating layer. The clouds are so tenuous that they are invisible to us when directly overhead.

Despite being so high, noctilucent clouds are also, like the much lower nacreous clouds, a form of wave cloud. They are generated by a gravity wave produced by mountains on the surface that has propagated up through the atmosphere. High in the upper atmosphere there is a tendency for the winds to carry the cloud particles towards the west or south-west, but the high-altitude waves themselves tend to remain stationary, giving structures that seem to move in the opposite direction.

## How do Noctilucent Clouds Form?

How these clouds form is actually a bit of a mystery. Meterologists are still unsure as to where the water vapour, and the particles (freezing nuclei, see page 54) onto which it freezes come from this great altitude. The particles may be dust from meteoric particles entering from space, or from clusters of ions created by cosmic rays. As for the water vapour, there is no easy mechanism by which this could be transported from the Earth's surface right up through the troposphere, stratosphere and mesosphere. Scientists now think it likely that the vapour is produced by the breakdown of methane gas, which can make its way to the top of the atmosphere.

# Aurorae

Strictly speaking, aurorae (in Britain, the *aurora borealis* or 'northern lights') are not meteorological phenomena, but these dramatic features are of course fascinating for anyone interested in looking at the sky. Aurorae occur far above any clouds (even noctilucent clouds), at altitudes between 100 and 1,000km (62 and 621 miles). They arise when energetic particles from the Sun cascade into the atmosphere in regions around the magnetic poles, and excite various molecules, causing them to radiate light.

In Britain, these stunning auroral displays are most commonly seen from Scotland, Orkney and Shetland, but strong displays may be visible much further south. On rare occasions when there is a major geomagnetic storm, aurorae have even been seen as far south as the Mediterranean and Caribbean.

The colours of the northern lights depend strongly on the particular gases involved. (They are also dependent on the observer's eyesight: some people are unable to see red auroral colours.) The most common colour in bright displays is 'auroral green', a distinctive green tint that derives from excited oxygen molecules, typically at heights of 90–150km (56–93 miles). A red oxygen emission occurs at greater altitudes. If the energetic particles reach low altitudes of 100km (62 miles) or less, a red nitrogen emission may occur.

Auroral displays are remarkably varied. They may be no more than a greenish glow, low on the horizon toward the north, or they may appear as patches, looking rather like misty clouds illuminated by distant artificial light. One distinctive form is an arch, highest towards the magnetic pole –towards the north, if you are in Britain – usually with a well-defined lower boundary and a more diffuse upper border. You can often see distinct rays radiating upwards from the arch. When rays are very long and reach heights of 1,000km (621 miles) or more, they may actually be in sunlight, which gives them a bluish-violet colour. At other times, wide ribbons of light appear, that may wave about – just like curtains blown by a wind. If a display moves overhead, it often shows a radiating structure, known as a corona (not be confused with the optical phenomenon of the same name, described in Chapter 5).

**Left** Aurorae at Tarbat Ness, Ross-shire.

**Below** A rainbow at St Michael's Mount, Cornwall.

# 5. Colours in the Sky and Other Optical Phenomena

The colour of the sky itself and of clouds, together with the numerous optical phenomena that may occasionally be visible, can give us information about current conditions. In a few cases they can also indicate how the weather may develop.

# Why is the Sky Blue?

That's a question that many children ask, but quite a few adults find hard to answer! Very broadly speaking, we perceive colour because light receptors within our eyes respond to different wavelengths in light, and then transmit information from our eyes to our brains as to what colour we are 'seeing'. We perceive the sky as blue because the molecules of nitrogen and oxygen (which make up most of air) are a size that scatters short-wave blue and violet light in all directions. But they have no effect on the longer wavelengths, such as red. Humans, unlike most insects, can't see violet light very well, so the sky appears blue to us rather than violet.

## Aerial Perspective

Water molecules in the atmosphere scatter white light and this will dilute the blue colour of the sky. At low levels, where water vapour is nearly always present, this causes aerial perspective, where distant objects appear less distinct and paler in colour. At high altitudes the air is dry, so the sky appears an intense, deep blue.

# Sunrise and Sunset

The phenomenon that results in red skies at sunset is called scattering. When the Sun is low on the horizon the light travels through more of the atmosphere towards us than it does during the day. Because of the nature of the molecules, more of the short-wave blue wavelengths are scattered aside than other wavelengths in the spectrum, and the light appears red or orange to us. The same process occurs at sunrise and sunset, which is why the rising or setting Sun and Moon always appear reddened. The same effect occurs whenever light takes a long path through the atmosphere before it reaches our eyes. For example, if a layer of cloud extends almost to the horizon, the distant strip of sky that remains visible just below it will frequently look orange or red.

At sunset, mountain peaks are sometimes lit up by a dramatic sequence of colours – yellow, pink, red and, finally, purple – as the Sun sinks lower. The same effect can happen, in reverse, at sunrise. This effect is known as 'Alpenglüh' ('Alpine glow' in German) and can also appear on high clouds, such as high cumulonimbus, which are on the opposite side of the sky to the Sun.

The sky itself goes through a sequence of colours in the twilight arch, the period just after sunset (and also sunrise). Shortly after sunset, the area closest to the horizon takes on a yellowish tint. Higher up, the sky is a pinkish shade, and still higher there is a blue area that gradually blends into dark blue above. Later on, the yellow area tends to turn to orange and the pink may grade into purple, and then into dark blue.

**Above left** Blue skies at Wastwater, Cumbria.

# Red Sky at Night

Most of the sayings linked to weather lore have little basis in fact, but an exception is the well-known British saying, 'Red sky at night, shepherds' delight'. Most weather systems cross Britain from west to east, so a red sky – clouds illuminated by red light – suggests that the sky is clear to the west, and the clouds overhead are passing away to the east. Similarly, 'Red sky in the morning, shepherds' warning', also has some truth. It suggests that although the sky may be clear in the east, clouds accompanying a depression are moving in from the west, bringing bad weather.

When the sky is clear to the horizon opposite the Sun, it is often possible to see the shadow of the Earth cast on the atmosphere. This appears as a dark blue-grey band along the horizon, with

**Left** Sunrise over the Marsden Moor Estate, West Yorkshire.

a reddish upper border. As the Sun sinks in the west, so the shadow of the Earth gradually darkens as it rises in the east, until it merges into the dark sky above. The effect is most noticeable when there is some haze or moisture in the atmosphere. There is a similar effect at sunrise, with the shadow sinking in the west as the Sun rises in the east. But because the atmosphere is often clearer in the morning, the shadow at sunrise tends to be less distinct and more difficult to see. If anyone is near the top of a mountain at sunrise or sunset they may be able to see the mountain's shadow appearing as a broad cone of darkness cast upon the atmosphere or on a layer of low cloud. Some mountain shadows have become well-known: the shadow of Mount Fuji in Japan is particularly famous and large numbers of people make the trek to the top just to see the shadow at sunrise. Mountain shadows always appear triangular, although the sides of the shadow are actually parallel and merely appear to converge because of perspective.

# The Purple Light and Other Colours

Unusual colours may, on occasion, transform the sky. One such is the striking and unmistakeable effect known as 'the purple light', which fills the sky with a vibrant purple shade. This rare effect arises only after a major volcanic eruption has ejected sulphur dioxide high into the stratosphere. The sulphur dioxide then combines with water to produce tiny droplets of sulphuric acid, which are of a size that scatters red light in all directions. The red light combines with the normal scattered blue light to produce the eerie purple glow. This dramatic coloration was seen back in 1991, with the violent eruption of Mount Pinatubo in the Philippines.

Volcanic eruptions can also be responsible for other colours, because violent eruptions may eject large quantities of dust and ash into the atmosphere. The twilight arch often broadens and the yellow and orange tints become more prominent. Frequently, the whole sky is covered in an orange glow. The dusty material generally occurs in a distinct layer, and undulations in the layer produce thicker and thinner regions that are visible as striations across the sky. Because of the direction of illumination, these usually appear roughly at right angles to the light from the Sun.

**Above right** Crepuscular rays emerge from a cloud over the sea.

# Crepuscular Rays

When the atmosphere is slightly hazy – perhaps with high humidity – rays of light separated by bands of shadow may be visible in the sky, often appearing to radiate outwards below clouds, or around the Sun. These crepuscular rays may be shafts of sunlight that, very frequently, penetrate between the individual cloudlets of stratocumulus cloud, or bands of shadow that radiate across the sky, cast by distant clouds or mountains that block the sunlight. When accompanied by bands of shadow, the latter are sometimes mistaken for shafts of rain. Crepuscular rays are sometimes known as 'the Sun drawing water', 'Jacob's Ladder', or (particularly to sailors) as 'Apollo's backstays'.

Sometimes the bands of shadow are so long that they converge at the antisolar point – the point on the sky directly opposite the Sun – when they are known as anticrepuscular rays.

# Optical Effects

The various optical effects that sometimes arise in the atmosphere fall into two broad categories: those caused by water droplets and those arising from ice crystals.

## Rainbows

We are all familiar with rainbows, often seen in showery weather when rain and sunshine are present at the same time. You can even see them in waterfalls or spray from a sprinkler system.

### Moon Bows

The Moon also creates rainbows, sometimes called moon bows, but these are naturally weaker and so the colours are not nearly as noticeable as those caused by the Sun.

To see a rainbow, you have to have the Sun behind you and the water droplets in front of you. The rainbow we usually see is the primary bow, with a radius

**Left**  A rainbow at Doyden Point and Kellan Head, Cornwall.

of approximately 42° (it is common to measure 'distances' in the sky by angle, from the edge of the object to your eye). Rainbows seen from the ground are always incomplete parts of a circle that is centred on the antisolar point – the point on the sky directly opposite the Sun. You will often notice that you can only see part of the rainbow, either because no rain is falling over the area that would produce part of the arc, or because that region is in the shadow cast by another cloud. Sometimes you may be lucky enough to see a complete circular rainbow from inside an aircraft.

Primary rainbows are caused when sunlight enters falling raindrops is reflected by the rear surface, and is dispersed into the spectral colours by refraction as the light passes through the drop and the different wavelengths are refracted by differing angles and 'split up'. You may have demonstrated this at school by directing light through a glass prism. The colours in the primary bow run from red on the outside of the bow to violet on the inside. Sometimes there are faint, pastel-coloured bands within the primary bow, usually pale violet to greenish. These are known as supernumerary bows (or sometimes as interference bows) and happen when the light has taken slightly different paths through the raindrops. The higher the Sun is in the sky, the lower the top of the rainbow, so if the elevation of the Sun at your position is greater than 42°, you won't see the primary rainbow at all. If the Sun is very low on the horizon, an almost perfect semicircular rainbow may appear, but it will look completely red in colour – all other wavelengths will have been scattered aside before they reach your eyes due to the Sun's position.

Secondary rainbows are fairly common as well, and are also centred on the antisolar point. Look out for their reversed sequence of colours, with red on the inside and violet at the outside 'top' of the arc. Appearing at a radius of approximately 52°, they are caused when sunlight is reflected twice within each raindrop.

The larger the raindrops, the brighter the bows. Large drops also produce very prominent red coloration in the bows, while with small droplets the red tint is less prominent, and the spacing between the supernumerary bows (if present) increases. With very tiny drops of rain, all coloration disappears completely to leave a white arc, otherwise known as a fogbow.

Dewdrops may also refract light into a spectrum of colours, creating a dewbow, like a faint rainbow lying on the ground. You will probably have seen the dewbows created by dewdrops held on spider's webs, or spun between blades of grass – especially in autumn. These bows are faint and often go unnoticed. Like rainbows, they are centred on the antisolar point but, because they are horizontal, have the shape of a partial ellipse or hyperbola. Occasionally you can see similar bows in tiny dewdrops resting on the surface film of ponds and pools of water.

## Alexander's Dark Band

When both primary and secondary bows are visible, you will notice the sky between them looks noticeably darker to you than it is elsewhere. This is because sunlight between the two bows has been reflected away from where you are observing. The region is known as 'Alexander's dark band', in honour of Alexander of Aphrodisias, who first noticed the effect in around 200AD.

# Glories, the Brocken Spectre, and the Heiligenschein

This fascinating group of effects may result in some striking phenomena. However, they are all easily explained. A glory – so named because it resembles (and may even be the origin of) the haloes that are shown around the heads of saints and other religious figures – is a set of coloured rings around the shadow of one's head. It occurs when the shadow is cast on a bank of mist or cloud. Like a primary rainbow, the red is on the outside, and comes from the Sun interacting with water droplets – in this case, usually in mist or clouds. But a glory is caused by diffraction of the light, rather than the reflection and refraction that cause rainbows. If there is a group of people, each observer will only see the glory surrounding the shadow of their own head. There may be multiple concentric rings, and the smaller the droplets of water, the greater the radii of the rings.

**Below** A glory, surrounding the shadow of the observer, cast onto misty cloud.

# Glories and Aircraft

Glories are also commonly seen from inside an aircraft flying above a layer of cloud. If the aircraft is close to the cloud layer, the plane's shadow is often distinctly visible in the centre of the glory.

A related effect is when your shadow is cast on mist or cloud (frequently accompanied by a glory) but seems to the magnified into a menacing presence. This is the famous Brocken Spectre (named after a mountain in Germany), actually an optical illusion related to an effect known as the tunnel illusion. It happens because our eyes have no way of being certain of our distance from the shadow we observe. Consequently, our brains try to interpret other clues. Its size seems to be accentuated if nearby objects – such as rocks – cast lines of shadow that to us appear to converge into the mist, because of perspective.

Apart from dewbows there is a second effect caused by dewdrops on the ground. A white halo, known as the heiligenschein (a halo or, literally, 'holy light') may appear round the shadow of an observer's head. The effect is most marked when drops of water rest on blades of grass, supported away from the surface by the tiny hairs that cover the grass. The blades of grass reflect light and the water droplets, together with the small air gap, concentrate the light in the direction of the observer. A similar, although fainter, effect may sometimes be seen in bright moonlight.

Another form of heiligenschein is the 'hot spot'. This optical effect may be seen on dry grass, fields of grain, and on the leaves of trees and similar objects. You can see it most clearly looking out from an aircraft, and it is well known to aerial photographers. In the case of an aircraft, a bright spot of light appears on the ground where the shadow of the plane would fall (if it were visible), and this spot moves with the aircraft. A similar effect may be seen on the ground when looking directly away from the Sun. When we look away from the Sun, only the brightly illuminated surfaces are visible to us, and they hide any shadows. This gives the appearance of a roughly circular bright spot of light.

## Coronae and Iridescence in Clouds

When one or more series of coloured rings appears around the Sun or Moon seen through thin cloud, this is a corona (Latin for 'crown'). You may need to hide the Sun behind some object to see its coronae, but coronae around the moon are often visible directly.

A visible corona implies that the cloud consists of water droplets; the effect is caused by diffraction. When you can see all the rings you will notice an inner area, known as an aureole, which is bluish-white with a brownish to reddish outer ring. Outside the aureole (which is not always visible) there is a set (or sets) of coloured rings, with violet on the inside and red on the outside. The purity of the colours depends on the droplet size, with uniform droplets giving the purest colours. When the droplets are a mixture of sizes, often only the aureole is visible. If the cloud is broken or ragged in outline, only parts of the rings may be visible.

**Above** A corona.

## The Solar Corona

There is a corona around the Sun that extends millions of kilometres into space. It is rarely visible to the unaided human eye, and more often observed and monitored through instruments. This is because the direct light is normally so strong that the effect is difficult to see. One exception is during a solar eclipse, but be sure to take the correct safety precautions to protect your sight. There is also a form of telescope called a coronograph that blocks direct light from the Sun's disk to make the corona safer to observe.

Iridescence is an effect of bright bands of colour that tend to run along the edges of clouds. As with coronae, the bands are caused by diffraction by water droplets and are strongest when the clouds are 30–35° away from the Sun. Normally red and green tints are visible, sometimes you can see yellow and, occasionally, blue. As with coronae, the strongest colours appear when there is little variation in the droplet size. Iridescence is responsible for the strong colours seen in nacreous clouds (see page 130), although in this case the cloud particles consist of ice, rather than water droplets.

# Ice Crystal Effects

There is a whole range of optical effects that may be visible on occasion, although many of these are extremely rare, and so cannot be described in detail here. Some effects may be visible only if the observer is at a particular latitude on Earth.

# Haloes

Probably the most common effect caused by ice crystals is a halo. This frequently occurs in the thin cirrostratus cloud that covers the sky ahead of an advancing depression so, along with related effects, is an indication of bad weather to come. Haloes are surprisingly common, but are noticed less than you might think. In fact, in Britain haloes are visible roughly every three days, because of the frequency with which depressions cross the country.

The most common halo is a ring of light around the Sun (or Moon) with a radius of 22°. (That's about the angle covered by the splayed fingers of one hand, held at arm's length.) To see the halo you may need to block out the Sun itself, positioning yourself so it is behind some object. The halo is usually white but may occasionally show very slight tinges of colour, with violet on the outside and red on the inside. Haloes round the Moon rarely show any colour, because light from the Moon is so much fainter than sunlight.

A much larger halo very occasionally occurs, similar but with a radius of 46°. It has the same colours as the 22° halo but, because it is so much weaker, it is rarely visible.

**Below** A 22° halo in thin cirrostratus cloud.

# Parhelia (Mock Suns)

Haloes are frequently accompanied by parhelia (the singular is 'parhelion'), also known as 'mock Suns' or 'Sun dogs'. These bright spots of light lie at the same altitude as the Sun. Depending on the exact altitude of the Sun, they may lie on the ring formed by the 22° halo, or slightly outside it. Parhelia often show bright, white 'tails' that point away from the Sun, and when they are bright also display spectral colours. The colours may be particularly striking, especially when a parhelion appears (without a halo) in a patch of cirriform cloud. A similar effect (a 'mock Moon') or paraselene occurs with light from the Moon, although its colours are rarely visible.

# Circumzenithal and Circumhorizontal Arcs

The most common, strongly coloured ice crystal effect is the circumzenithal arc. This consists of a portion of a circle centred on the zenith (the point directly overhead). It is normally about 120° long, and is symmetrical about the line joining the zenith and the Sun. It will touch the 46° halo (if present) at a point immediately above the Sun. Circumzenithal arcs are very striking and are often described as 'upside-down rainbows'. However, they are not related to actual rainbows, being caused by ice crystals rather than raindrops.

The circumhorizontal arc is perhaps the most spectacularly coloured phenomenon. This is a brilliant arc that lies parallel to the horizon. It may only be seen when the elevation of the Sun is greater than 58°, so is generally only observable from low latitudes.

**Right** A bright circumzenithal arc, seen high in the sky above a chimneytop.

# Other Ice Crystal Phenomena

There are numerous optical effects that arise from the passage of light through ice crystals. Here are a few others that occur quite frequently but haven't yet been covered.

**Parhelic circle** A white arc, at the same altitude as the Sun and parallel to the horizon. A complete 360° circle is extremely rare.

**Sun pillar** A vertical pillar of light above and below the Sun. This appears when sunlight is reflected by the surfaces of ice crystals shaped as flat plates, floating horizontally in the air.

**Subsun** An elliptical white patch of light, taller on the vertical axis, that appears below the location of the Sun. You can only see it if you are in an aircraft or high on a mountain. The centre of the subsun is the same distance below the horizon as the distance of the Sun is above it. It is quite common to see this when an aircraft is flying in cirriform (ice crystal) cloud. On very rare occasions, additional bright spots of light may appear on each side of a subsun, roughly 22° away from it. These spots of light (known as 'subparhelia' or 'subsun dogs') are similar to the more familiar parhelia.

# Refraction

There are a number of effects that arise through refraction in the atmosphere – where light is bent through an angle when it encounters air of different density. The denser the air, the greater the deviation of the rays of light.

# The Sun

The image of the Sun (in particular) is frequently distorted by the presence of various layers of air at different densities. The image may be broken up into strips, or appear somewhat like a vase with a distinct 'foot'. Occasionally, this 'foot' on the horizon seems to expand sideways, and then the outline of the Sun appears like the Greek letter omega. For obvious reasons, these are know as the 'Vase' and 'Omega' effects, respectively.

Another refraction effect is the green flash, where the last remaining portion of the Sun at sunset turns green. A similar effect occurs right after sunrise, but is less often seen and reported, partly because of the difficulty of knowing exactly where the Sun will appear. These are most often observed in clear air, when there is very little scattering of light and so most reaches the observer. The green flash effect is best seen over a sea horizon, where a larger green segment may sometimes be visible. Sightings are often made by aircraft pilots, who are able to observe the unobstructed horizon. Even more rarely, at sunset the last portion of the Sun may appear blue.

## The Sun's Refracted Image

When the sun is setting, or rising, rays are refracted by the atmosphere and make it appear higher in the sky than it actually is. At sunset it may look to you as if the bottom of the Sun is touching the horizon. In fact, the whole Sun is below the horizon but its image is 'raised' by refraction in the atmosphere. The same effect happens with the Moon.

**Above** Sunset over Lizard Point and Kynance Cove, Cornwall

# Distant Objects

Air density decreases upwards, so images that reach us from distant objects are subject to the effects of refraction, four of which are given here with their cause:

**Looming** An abnormally large refraction of the object makes it appear much higher. Distant objects that are normally unseen, because they are below the horizon, become visible to the observer. This is a result of air density decreasing rapidly with height.

**Sinking** This is the opposite of looming. Objects that are above the horizon, and so should be visible to the observer, are depressed below their normal position and may become invisible. This is caused because air density decreases more slowly than normal, or it may even increase with height.

**Stooping** The shape of the objects appears to be compressed vertically. This is caused when the lowest layer of air is warmer and has lower density than normal.

**Towering** Objects appear elongated vertically. This results when air density decreases more rapidly with height than normal.

# Mirages

Mirages also involve abnormal refraction by layers of air of different densities in the atmosphere. There are two broad classes of mirages. In one, generally referred to as a 'superior' mirage, the image of a distant object appears to be in 'the wrong place': it appears in a different location to the position it occupies under 'normal' conditions, when seen through a layer with uniform density. The other, 'inferior' mirage is a more commonly seen form, where what appear to be pools of water seem to be lying on a hot road. The 'water' is actually an image of the sky. The illusion of 'water' is because rays of light from the sky have been sharply curved by the layer of hot, low-density air that lies immediately above the road.

A superior mirage occurs when a layer of warm air lies above a cold surface. This may happen over an expanse of ice or snow, but is most common over the sea, and especially in spring, when the water temperature is low. Distant objects appear to be floating in the air and, may sometimes appear to be inverted (upside down). Multiple images, some inverted, often appear. Frequently, the layers of air with different densities are shallow, so the distant images may are seen as narrow strips and are difficult to distinguish.

## Fata Morgana

In one notable effect, known as Fata Morgana, the images of distant objects are elongated, and can look like vertical cliff faces or groups of distant buildings, mysteriously suspended over the sea. The name comes from the Italian for Morgan le Fay, the Arthurian sorceress. Italian folklore has her living in a floating magical castle over Mount Etna, and this mirage effect is also common off the shores of Sicily. In reality, the mirage images are highly distorted images of the sea or distant ice floes.

**Above** Fata Morgana seen off the North Sea coast.

# Weather Forecasts

Weather forecasts are available in so many different media nowadays that it can be difficult to decide which are most trustworthy, or most suitable for your needs. You can download specialised weather apps for almost any situation, including forecasting apps for mountain areas, inshore waters, or for severe weather warnings. There also forecasts for pollen counts, and warnings of severe heat or cold, both routinely included in many television forecasts, and for commuting or farming conditions. Some of the latest smartphone and tablet weather apps are extremely accurate in showing the location and duration of rain (even showing the onset or cessation times to the nearest minute) although tracking the intensity of rainfall remains uncertain.

Although television weather forecasts have improved greatly in recent years and are generally suitable for most purposes, some of the 'improvements' are designed to make the forecast accessible to all, but are sometimes frustrating for those who want more detail and precision. There has been, for example, a general trend to omit isobars from television forecasts, replacing these with generalised wind arrows, often animated. But for anyone with a slightly greater interest and knowledge, isobaric charts offer valuable information.

# Data Collection

All modern forecasts use a method known as numerical weather prediction (NWP), based on the use of supercomputers, which are fed the raw observational data from a vast number of sources. These observations, worldwide, are obtained simultaneously at specific times. These times are laid down by the World Meteorological Organisation (WMO) and are always specified in Coordinated Universal Time (UTC). UTC is the time on the Greenwich Meridian, and doesn't take into account adjustments for Summer Time (Daylight Saving Time). The standard synoptic times are 00:00, 06:00, 12:00 and 18:00 UTC, but most major meteorological stations also make observations at the intermediate times of 03:00, 09:00, 15:00 and 21:00 UTC. Observations are made available, worldwide, via the Global Data-Processing System, coordinated by the WMO.

## Sources of Data

There are many sources for observations. Some data comes from traditional types of observation, where people take readings from a whole suite of instruments at specific times, either at fixed land stations or aboard ships at sea. Most comes from automated weather stations on land, often sited at virtually inaccessible locations, or from automated weather buoys, some even anchored in the mid-Pacific, or from aircraft. There also drifting buoys and automated floats that sink to pre-determined depths and periodically return to the surface to transmit their data, via satellites. These are primarily devoted to oceanography and climatic studies, but some of their data, such as sea-surface temperatures, also has relevance for weather forecasting. Modern instrumentation is becoming increasingly sophisticated: it is now possible to obtain information on wind speeds and directions directly above ground stations by using Doppler wind profilers (which use radar), right up to the altitude of the tropopause.

Data from aircraft in flight are of considerable use, but the observations are limited to specific flight levels. Radiosondes, instruments carried aloft by balloons and transmitting data by radio, can provide profiles of conditions through a considerable depth of the atmosphere (to about 20km [12½ miles]), and are released at specific times (worldwide) twice a day. There are even systems carried by ships that are programmed to inflate and release balloons automatically without human intervention. Releases from the surface occur about 45 minutes before the nominal observation times (00:00 and 12:00 UTC) to ensure time for the balloons to lift the instruments to suitable altitudes and that the observations are comparable to those obtained by other methods. Modern radiosondes are typically fitted with GPS receivers to determine their exact position and normally measure altitude, pressure, temperature, relative humidity, wind speed and wind direction. Some are able to determine cosmic rays (high energy radiation) and others (known as ozonesondes) can determine ozone concentrations.

Images from meteorological satellites are commonly shown in television forecasts, but the same images also provide data for meteorologists to differentiate different cloud heights and types. As well as images, many other streams of meteorological data are provided by both specifically meteorological satellites and Earth-observation satellites. There are two categories of meteorological satellites: polar-orbiting and geostationary. (The majority of

specifically Earth-observation satellites are polar-orbiting.) Both types offer particular advantages, and both are extremely important in forecasting. Their instrumentation is becoming increasingly sophisticated and significant. Recent developments in determining atmospheric conditions from orbit – so-called 'topside sounding' – even include methods for determining surface pressure from orbit.

Polar-orbiting satellites (at typical altitudes of 700–800km [435–500 miles]) are normally placed in what are termed 'sun-synchronous' orbits, so that they pass over a single swathe of the Earth at the same time twice each day, once on a north–south pass and once on a south–north pass. Current polar-orbiters include the NOAA series (launched by the USA's National Oceanic and Atmospheric Administration) and METOP-A and METOP-B, operated by Eumetsat, (the European Organisation for the Exploitation of Meteorological Satellites). There are also comparable satellites operated by Russia, China and India.

The animated sequences of the motion of depressions and the changes in cloud cover shown in television forecasts are generally drawn from the images returned by geostationary satellites. These are geosynchronous satellites, which operate in a much higher orbit, 35,786km (22,236 miles) above the equator, and complete one orbit in the same time as the Earth completes one rotation. This means that they remain, essentially, stationary over one point of the equator. This makes them ideally placed to provide continuous coverage of the Earth's surface, although the curvature of the Earth places limits on the area that may be covered by a single satellite (approximately 120°, rather than a full 180°) and their distance above the Earth means that the resolution is lower than that obtained by polar-orbiters, in their much lower orbits. There are a number of such geostationary meteorological satellites, collectively providing continuous coverage around the Earth. These include GOES-E and GOES-W (NOAA satellites), Meteosat (Eumetsat), Hinawari (Japan), Fengyun (China), and Indsat (India).

Radiosonde balloons typically lift instrumentation to heights of 20–21km (12½–13 miles), but the lowest satellite orbits are at altitudes of around 700km (435 miles), far above this height. Sounding rockets are able to bridge the difference in altitude, but can provide measurements only for a single location and at one specific time. This means there is an extremely significant lack of knowledge about conditions in the upper atmosphere, and it is difficult to determine what implications they may have for weather close to the surface of the Earth.

Scientists track solar cycles and activity by counting sunspots, cooler patches that appear on the Sun's surface, and appear darker. For many years meteorologists have suspected that solar activity has an effect on the weather, and there have been numerous attempts to match sunspot numbers to meteorological conditions. Since sunspot numbers are only an indirect measure of solar activity, attempts to find a direct link to meteorological conditions have so far failed.

## Solar Effects on the Weather

Recently, scientists have found that solar activity and upper-atmosphere conditions are related to the strength of the polar vortex (see page 29), and that could mean that they thus affect both the jet stream and conditions close to the surface. There are now several projects to create high-flying, long-duration drones to obtain observations from great heights and discover more. (Meteorological drones are already used for special research projects at near-Earth altitudes.)

# Creating Forecasts

Because observations are obtained at the same specific time intervals worldwide, they can be used to assess the overall global meteorological conditions at a specific time. The data may be used to create synoptic charts or for numerical weather forecasting. The latter process uses models of the atmosphere based on a fixed grid of points. The number of grid points used for the model is dependent on individual meteorological services and their supercomputing capacity. In the case of the UK Met Office, one global model uses a grid with a resolution of 17km (10½ miles), and various local models have grids down to a resolution of just 1.5km (almost a mile). Furthermore, calculations are undertaken at multiple levels in the atmosphere (70 in one of the Met Office models), which generally includes all of the troposphere, the area in which most significant weather takes place. Using equations that describe the physical behaviour of the various parameters (temperature, pressure, etc.) the supercomputers carry out trillions of calculations to determine the likely situation in the atmosphere at particular times in the future. This modelling forms the basis for the majority of forecasts.

# Analysis Charts

Publicly available forecasts often take the form of an 'analysis chart', which is produced from the synoptic data, and also indicates the location of fronts. The exact location of fronts is often difficult to determine by relying on observational data; as previously mentioned, fronts are not a sharp boundary at the surface, but rather a frontal surface where mixing of two air masses occurs, and they may be as much as 200km (124 miles) in width. However, analysis charts such as the one below, from the UK Met Office, give an accurate picture of the pressure distribution and the accompanying isobars.

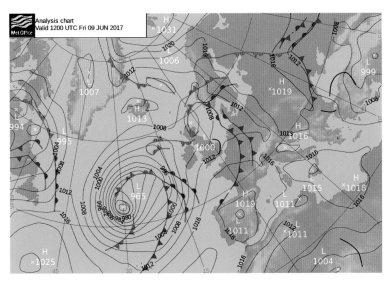

This particular chart shows the situation at 12:00 UTC (i.e., noon). Similar charts are often shown for 00:00 UTC (midnight). This chart – Friday, 09 June 2017 – wasn't chosen to illustrate any specific meteorological situation but has some interesting features, nevertheless: note the closely spaced isobars (indicative of high wind speeds) on the western side of the depression centre in the mid-Atlantic Ocean, which is marked by the small white cross and had a pressure of 965hPa. A lesser, indistinct low (1,000hPa central pressure) is located just off north-eastern Scotland.

# Forecast Charts

Analysis charts are generally accompanied by forecast charts, which usually illustrate the forecast at 12-hour intervals, in advance. The next chart shows the forecast for 24 hours after the analysis chart (see 'T + 24' in the top left corner).

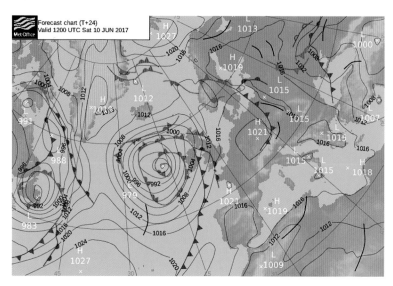

From this chart you can see that the forecast is for the depression out in the Atlantic to fill slightly (from 965 to 979hPa) as it moves closer to Ireland. The spacing of the isobars on the western side is now greater, indicating that the wind speeds are expected to decrease slightly. The low formerly off the north-east of Scotland is expected to dissipate and no longer appear distinct, as will one of the warm fronts from the main depression. The other warm front is forecast to run roughly north from Devon up along the border between Wales and England.

This chart also illustrates the situation, known as a col, where an area of slack pressure exists between two low-pressure regions (depressions) and two high-pressure regions (anticyclones). In this case there are the two depressions, one in the mid-Atlantic and the other developing over Newfoundland, with the Azores High to the south, and a smaller area of high pressure over southern Greenland. Cols tend to be short-lived, because even the slightest change in the pressure pattern may cause a substantial change in their position.

The original analysis chart for Friday, 09 June 2017 was accompanied by a series of forecast charts at 12-hour intervals, extending four days ahead. The chart below is the forecast chart for noon on Tuesday, 13 June 2017 – 96 hours ahead of the analysis chart. From this chart you can see that the depression that was originally in the mid-Atlantic is expected by then to have passed right over the British Isles with the remnants, with central pressure of 991hPa, located over Finland.

A ridge of high pressure has extended from the Azores High along the English Channel, with highest pressure of 1019hPa over the Straits of Dover. The cold front trailing behind the original depression is forecast to curve south from the Finland–Russia border over eastern Europe, run along the northern side of the Alps, and then cross France and lie over the Bay of Biscay. There is an occluded front over Finland. The slightly weaker depression (981hPa) with various frontal systems and low-pressure troughs (indicated by solid black lines), previously over Newfoundland, is expected now to lie in the mid-Atlantic.

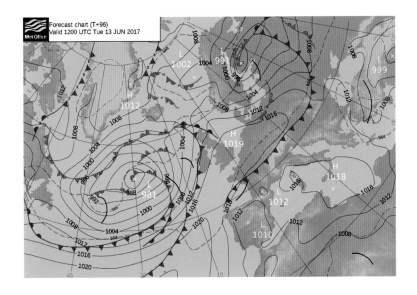

Comparing a forecast chart with a later
analysis chart for the same time, as below,
can illustrate how close forecasts may be
(and also where deviations occur).

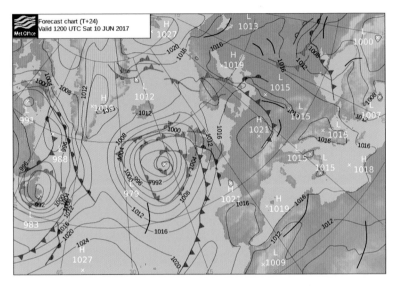

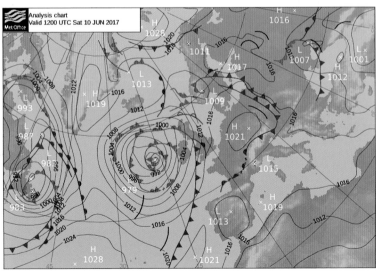

By comparing these two charts you can see that the forecast was correct for the position of the centre of the main depression and its central pressure (979hPa on both charts), as well as the location and pressure (1021hPa) of the anticyclonic region over Germany. The central pressure and location of the depression over Newfoundland are also as predicted. Although there are some slight differences, the overall pattern of fronts is similar on both charts.

As mentioned previously, the precise motion of frontal systems is complex, being partly governed by the nature of the surface and the resulting friction. In the case of cold fronts, friction with the surface (particularly over the land) may cause the initially backward-sloping front to steepen – because friction decreases with increasing altitude. Or cold air at altitude may even overrun the surface front, giving a 'nose' of cold air aloft, ahead of the surface frontal zone.

# Nowcasting

Television weather forecasts (particularly forecasts for local areas) often use 'nowcasting', a term used for the preparation of weather forecasts for a short period ahead – frequently 6 hours. Nowcasting is also often used to produce very short-term forecasts – perhaps no more than 30 minutes ahead – for specific purposes, such as forecasting conditions at an airport.

Such forecasts often include a description of current weather (or the situation a short time ago), based on observations available in real time, rather than those obtained at the fixed, synoptic times. Nowcasts frequently include radar images of current rainfall, sometimes accompanied (when appropriate) with images obtained from lightning detection systems. These images are obtained from radio atmospheric signals emitted from lightning, known as 'sferics' (from 'atmospherics'). Both radar and sferics data can provide information about the location and severity of medium-scale weather systems that cannot be adequately determined from the fixed locations of standard weather stations.

# Further Reading

Dunlop, Storm, *A Dictionary of Weather* (2nd edition, Oxford University Press, 2008).

Dunlop, Storm, *Meteorology Manual: The Practical Guide to the Weather* (J.H. Haynes and Co., 2014).

Dunlop, Storm, *Weather: A Short Introduction* (Oxford University Press, 2017).

Hamblyn, Richard in association with the Met Office, *The Cloud Book: How to Understand the Skies*, (David & Charles, 2008).

Hamblyn, Richard in association with the Met Office, *Extraordinary Clouds* (David & Charles, 2009).

Kington, John, *Climate and Weather* (HarperCollins, 2010).

Met Office, Factsheets 1–19 (pdfs downloadable from: http://www.metoffice.gov.uk/learning/library/publications/factsheets

Watts, Alan, *Instant Weather Forecasting* (Adlard Coles Nautical, 3rd revised edition, 2007).

Watts, Alan, *Instant Wind Forecasting* (Adlard Coles Nautical, 3rd revised edition, 2010).

Watts, Alan, *The Weather Handbook* (3rd edition, Adlard Coles Nautical, 2014).

Williams, Jack, *The AMS Weather Book: The Ultimate Guide to America's Weather* (Univ. Chicago Press, 2009)

# Internet Links

## Current Weather

**AccuWeather**
www.accuweather.com/
**UK:** www.accuweather.com/ukie/index.
asp?

**Australian Weather News**
www.australianweathernews.com/
**UK station plots** www.
australianweathernews.com/sitepages/
charts/611_United_Kingdom.shtml

**BBC Weather**
www.bbc.co.uk/weather

**CNN Weather**
www.cnn.com/WEATHER/index.html

**Intellicast** intellicast.com/

**ITV Weather**
www.itv-weather.co.uk/

**Unisys Weather**
weather.unisys.com/

**UK Meteorological Office**
www.metoffice.gov.uk
**Forecasts** www.metoffice.gov.uk/
weather/uk/uk_forecast_weather.html
**Weather Observation Website** wow.
metoffice.gov.uk

**Surface pressure charts**
www.metoffice.gov.uk/public/weather/
surface-pressure/
**Explanation of symbols on
pressure charts**
www.metoffice.gov.uk/guide/weather/
symbols#pressure-symbols
**Synoptic and climate stations
(interactive map)**
www.metoffice.gov.uk/public/weather/
climate-network/#?tab=climateNetwork
**Weather on the Web**
wow.metoffice.gov.uk/

**The Weather Channel** www.weather.
com/twc/homepage.twc

**Weather Underground**
www.wunderground.com

**Wetterzentrale** www.wetterzentrale.de/
pics/Rgbsyn.gif

**Wetter3 (German site with global
information)** www.wetter3.de
**UK Met Office chart archive**
www.wetter3.de/Archiv/archiv_ukmet.
html

# General Information

**Atmospheric Optics**
www.atoptics.co.uk/

**Hurricane Zone Net**
www.hurricanezone.net/

**National Climate Data Centre**
www.ncdc.noaa.gov/
**Extremes** www.ncdc.noaa.gov/extremes/us-climate-extremes/

**National Hurricane Center**
www.nhc.noaa.gov/

**Reading University (Roger Brugge)** www.met.reading.ac.uk/~brugge/index.html

**UK Weather Information**
www.weather.org.uk/

**Unisys Hurricane Data**
weather.unisys.com/hurricane/atlantic/index.html

**WorldClimate** www.worldclimate.com/

# Meteorological Offices, Agencies and Organisations

**Environment Canada**
weather.gc.ca/canada_e.html

**European Centre for Medium-Range Weather Forecasting (ECMWF)**
www.ecmwf.int

**European Meteorological Satellite Organisation**
www.eumetsat.int/website/home/index.html

**Intergovernmental Panel on Climate Change**
www.ipcc.ch

**National Oceanic and Atmospheric Administration (NOAA)**
www.noaa.gov/

**National Weather Service (NWS)** www.nws.noaa.gov/

**UK Meteorological Office**
www.metoffice.gov.uk

**World Meteorological Organisation**
www.wmo.int/pages/index_en.html

# Satellite Images

**Eumetsat** www.eumetsat.de/
**Image library**
www.eumetsat.int/website/home/Images/ImageLibrary/index.html

**Group for Earth Observation (GEO)**
www.geo-web.org.uk/

**University of Dundee**
www.sat.dundee.ac.uk/

# Societies

**American Meteorological Society**
www.ametsoc.org/AMS

**Australian Meteorological and Oceanographic Society** www.amos.org.au

**Canadian Meteorological and Oceanographic Society: Climatological Observers Link (COL)**
www.colweather.org.uk/index.php

**European Meteorological Society**
www.emetsoc.org/

**Irish Meteorological Society**
www.irishmetsociety.org

**National Weather Association, USA**
www.nwas.org/

**New Zealand Meteorological Society**
www.metsoc.org.nz/

**Royal Meteorological Society**
www.rmets.org

**TORRO: Tornado and Storm Research Organisation** torro.org.uk

# Glossary

**air mass**
A body of air with specific characteristics (such as temperature and humidity), acquired when it remains over an area for some time. It generally retains its temperature and humidity (at least initially) when it moves away from the source area, and largely determines the weather of any region it crosses.

**anabatic**
Moving upwards. The term is typically applied to winds (such as a valley wind) or to the air at frontal systems.

**anticyclone**
A high-pressure region: a source of air that subsides from higher altitudes, and from which air flows out over the surrounding area. The circulation around an anticyclone is clockwise in the northern hemisphere.

**anticyclonic**
Moving or curving in the same direction as air circulating around an anticyclone (clockwise in the northern hemisphere, anticlockwise in the southern).

**antisolar point**
The point on the sky directly opposite the position of the Sun.

**backing**
An anticlockwise change in the wind direction, e.g., from west, through south, to east.

**Beaufort scale**
A numerical scale for the description of wind speed, ranging from 0: calm; 1: 1–3 knots (0.3–1.5m/s or about 1–3mph) to 12: above 64 knots (above 33m/s or above about 73mph).

**Coriolis force**
The apparent force, caused by the rotation of the Earth, that deflects any moving object (such as a parcel of air) away from a straight-line path. In the northern hemisphere it acts towards the right, and in the southern, to the left. It increases in proportion to the velocity of the moving object.

**cyclone**
A system in which air circulates around a low-pressure core, with two distinct meanings: 1) a 'tropical cyclone', a self-sustaining tropical storm, also known as a hurricane or typhoon; 2) an 'extratropical cyclone' or depression, a low-pressure area, which is one of the principal weather systems in temperate regions.

**cyclonic**
Moving or curving in the same direction as air that flows around a cyclone (anticlockwise in the northern hemisphere, clockwise in the southern).

**depression**
The most frequently used term for a low-pressure area. Air flows into a depresssion and rises in its centre. Known technically as an 'extratropical cyclone'. The wind circulation around a depression is cyclonic (anticlockwise in the northern hemisphere).

**dewpoint**
The temperature at which a particular parcel of air, with a specific humidity, will reach saturation. At the dewpoint, water vapour will begin to condense into droplets, giving rise to a cloud, mist or fog, or depositing dew on the ground.

**instability**
The condition under which a parcel of air, if displaced upwards or downwards, tends to continue (or even accelerate) its motion. The opposite is stability.

**jet stream**
A narrow band of high-speed winds that lies close to a breaks in the level of the tropopause, with two main jet streams (the polar-front and subtropical jet streams) in each hemisphere. Other jet streams exist in the tropics and at higher altitudes.

**katabatic**
Moving downwards. Used primarily in connection with katabatic winds (fall winds) that sweep down from high ground, and are

normally initiated by low temperatures over the higher ground. The term is also applied to the motion of the air at certain frontal systems.

## latent heat
The heat that is released when water vapour condenses into droplets or freezes into ice crystals. It is the heat that was originally required for the process of evaporation or melting.

## occluded front
A front in a depression system, where the warm air has been lifted away from the surface, having been undercut by cold air. The front may, however, remain a significant source of cloud and precipitation.

## stability
The condition under which a parcel of air, if displaced upwards or downwards, tends to return to its original position rather than continuing its motion.

## stratosphere
The second major atmospheric layer from the ground, in which temperature initially remains constant, but then increases with height. It lies between the troposphere and the mesosphere, with lower and upper boundaries of approximately 8–20km (5–12½ miles), depending on latitude, and 50km (31 miles) respectively.

## supercooling
The conditions under which water may exist in a liquid state, despite being at a temperature below its nominal freezing point, 0°C (32°F). This occurs frequently in the atmosphere, often in the absence of suitable freezing nuclei. Supercooled water freezes spontaneously at a temperature of -40°C (-40°F).

## synoptic chart
A chart showing the values of a given property (such as temperature, pressure, humidity, etc.) prevailing at different observing sites at a single, specific time.

## synoptic scale
Weather phenomena that are approximately 200–2,000km (124–1,243 miles) across, thus lying between mesoscale and planetary scale phenomena in size.

## thermal
A rising bubble of air that has broken away from the heated surface of the ground. Depending on circumstances, a thermal may rise until it reaches the condensation level, at which point its water vapour will condense into droplets, giving rise to a cloud.

## tropopause
The inversion that separates the troposphere from the overlying stratosphere. Its altitude varies from approximately 8km (5 miles) at the poles to 18–20km ($11^1/_5$ –12½ miles) over the equator.

## troposphere
The lowest region of the atmosphere in which most of the weather and clouds occur. Within it, there is an overall decline in temperature with height.

## trough
An elongated extension of an area of low pressure, which results in a set of approximately 'V'-shaped isobars, pointing away from the centre of the low.

## veering
A clockwise change in the wind direction, i.e. from east, through south, to west.

## virga
Trails of precipitation (as ice crystals or raindrops) from clouds that do not reach the ground, melting and evaporating in the drier air between the cloud and the surface.

## wind shear
A change in wind direction or strength with a change of position. If, for example, the wind strength increases with increasing height, this is defined as vertical wind shear. If the wind strength changes with motion at a particular level, this is known as horizontal wind shear.

# Index

# Picture credits